Do Butterflies Bite?

Do Butterflies Bite?

Fascinating Answers to Questions about Butterflies and Moths

Hazel Davies and Carol A. Butler

With illustrations by William H. Howe

Rutgers University Press
NEW BRUNSWICK, NEW JERSEY, AND LONDON

Library of Congress Cataloging-in-Publication Data

Davies, Hazel, 1968–
    Do butterflies bite? : fascinating answers to questions about butter-
flies and moths / Hazel Davies and Carol A. Butler; with illustrations by
William H. Howe.
        p. cm.
    Includes bibliographical references and index.
    ISBN 978-0-8135-4268-3 (pbk. : alk. paper)
    1. Butterflies—Miscellanea. 2. Moths—Miscellanea. I. Butler, Carol A.,
1943– II. Title.
    QL544.2.D38 2008
    595.78'9—dc22
                                                                    2007029663

A British Cataloging-in-Publication record for this book is available
from the British Library.

Visit our website: http://rutgerspress.rutgers.edu

Manufactured in the United States of America

19.95

Contents

NINE Outdoor Butterflies 117

TEN Indoor Butterflies 145

Appendices 155

A color insert follows page 96

Preface

Walk into any bookstore's natural history section and you will find dozens of different field guides and personal, anecdotal tales recounting the experience of discovering butterflies and moths. What you will not find is one comprehensive book that fills the need for answers about these truly fascinating creatures. How fast do butterflies fly? Does a butterfly have ears? What are butterfly wings made of? We hope that this book fills this void as we try to answer all the questions that we have been asked over the years, and even those questions that we've asked ourselves.

In order to interest readers on all levels, we have tried to provide basic information as well as a lot of the most up-to-date scientific observations, experiments, and theories. We have worked our way through the scientific literature and tried to keep up with the news in the field, and we have had the enthusiastic support of several entomologists who directed us to the latest discoveries and gently corrected our errors. Fundamental questions pertaining to butterfly biology and behavior remain unanswered and drive current research, resulting in a stream of new discoveries and insights that will undoubtedly make some of our answers less than completely accurate.

Do Butterflies Bite? explains many aspects of the intriguing world of these amazingly diverse animals, from body structure and behavior to the dangers they face and their place in the environment. We have tried to write in a nontechnical, friendly style, within an easy to use Q and A format that we hope people will find comprehensive but accessible. Simple answers are provided immediately, and more details and examples drawn from research and theory follow for those whose interest has been aroused.

No matter where we go, we're always looking for butterflies—outdoors or indoors in butterfly exhibits. With patience, it is possible to get as close to them as you wish, and we both have traveled to many places to have this experience. Observing and photographing butterflies in their natural environments has enhanced our awareness and appreciation of their incredible behavior, variety, and beauty. In this book we provide some ideas on where you can go butterfly watching and some tips on how to successfully photograph your sightings.

We both live and work in Manhattan, in the heart of New York City, and we know that to urban dwellers the natural world can often seem quite far away. But public butterfly exhibits are now open in many cities around the world, and they make the wonders of nature and the pleasures of learning about insects accessible to everyone. You can find a comprehensive list of these indoor conservatories, exhibits, and public gardens at the end of this book. We have listed approximately 200 locations, with the latest contact information; however, we suggest you call, email, or check the facility's website before visiting, as some of the gardens are seasonal or may no longer be open by the time this book reaches you.

The photographs and illustrations throughout serve to enhance the text. Though it would be impossible to show you all the butterflies and moths we discuss in this book, their images are easily available through simple web searches using an Internet search engine that includes images.

We hope this book will convey our fascination with the complex lives of butterflies and moths and will stimulate your interest to go butterfly watching and learn more about butterfly conservation and their natural habitats.

Acknowledgments

Many people contributed to the completion of this book, and the authors greatly appreciate their gracious donation of time, help, wisdom, and advice. We are incredibly grateful to Bill Howe for allowing us unlimited use of his beautiful drawings. Many thanks to Michael Weissmann of Kallima Consultants for reviewing the entire manuscript. Your advice, constructive criticism, and enthusiasm for the project have undoubtedly benefited the final form of the book. And to Mark Deering of Sophia Sachs Butterfly House, for reading large portions of the early manuscript.

We are indebted to the various professionals who graciously spent time answering questions. Special thanks to the Netherlands contingent, particularly to Menno Schuilthuizen for his helpful suggestions on parts of the manuscript, and to Menno and Rienk de Jong of the National Museum of Natural History at Leiden for making it possible to include the photographs from their collection in the book. Thanks to Joop van Loon at Wageningen University for guiding us to two wonderful photographs, and to Yde Jongema at Wageningen University and Ko Veltman at Artis Zoo for their generosity and responsiveness. Thanks too to Erich Stadler at the University of Basel for allowing us to use his magnificent tarsus photograph.

We especially appreciate the generous donation of hours and hours of time by Anna Yovu, who worked tirelessly on compiling the conservatory list. Thanks also to Anna for continual support with the entire project, helpful suggestions, contributing questions, research, and advice on photograph selection. Thanks to all our butterfly colleagues for contributing questions and sharing their observations and fascination with butterflies day after day. Special thanks are due to Barbara Griffing, for valuable advice during the preparation of the initial proposal.

To Zane and Maggie Greathouse, and the staff at Greathouse Butterfly Farm in Earlton, Florida, thank you for your warm hospitality, for providing wonderful photographic opportunities, and especially for the delicious bag of Chinese honey oranges to take away. Thanks are also due to Joris Brinkerhoff and the staff at The Butterfly Farm in Alajuela, Costa Rica, for a most informative visit, the chance to photograph beautiful butterflies, and my first traditional Costa Rican meal. Thanks also to Robert Goodden, of World Wide Butterflies, for information on the history of conservatories.

Special thanks to Doreen Valentine, our editor at Rutgers University Press, for invaluable help, and to our agent, Deirdre Mullane, of The Spieler Agency, who envisioned the book that the early proposal could become and never wavered. We are grateful to Derik Shelor for his interest and for his thoughtful copy editing.

From Hazel, for encouragement, friendship, editing, and comments on the manuscript, thanks to Amanda Accamando, Sheridan Vichie, and Rachel Lambert—thanks for the emergency chocolate supplies just at the right time. To my wonderfully supportive family, thanks for always being there and for sparking and developing my early interest in nature. I am grateful for all the time I spent exploring outdoors with each of you.

From Carol, special thanks to Paul Goldstein of the McGuire Center for allowing me to roam around in a collection for the first time, to Don Davis for his speedy responses to my queries, and to David Wagner for teaching me about frass in Mission, Texas.

And especially to Andrew Davies and Aisha Butler for invaluable technical support, encouragement, and never-ending patience, our heartfelt thanks.

Do Butterflies Bite?

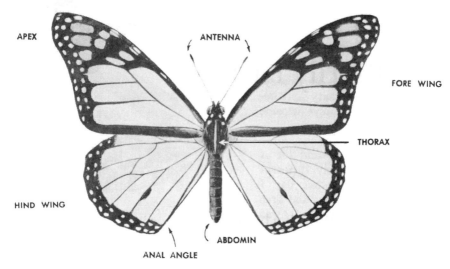

Figure 1. Morphological structures of a Monarch butterfly (*Danaus plexippus*). (*Drawing courtesy of William H. Howe, from* Our Butterflies and Moths, *page 25*)

Question 2: Is it a butterfly or is it a moth?

Answer: Butterflies and moths are very closely related, and throughout this book when we talk about butterflies, we will almost always be referring to moths as well.

Most people believe that moths fly at night and are small and dull in color, while butterflies fly during the day and are brightly colored. Although this is generally true, some moths are actually quite large and colorful and many butterflies are quite small and dull. Also, many species of moths are active during the day (*diurnal*), such as the African Peach moth (*Egybolis vaillantina*), which is colored a beautiful iridescent blue and orange.

The most reliable way to determine whether you are seeing a moth or a butterfly is to look at the antennae. Although there are exceptions, the antennae of a butterfly are usually straight and slightly thickened or clubbed at the tip. Many moths have feathery, toothed, or bristle-like antennae that come to a point

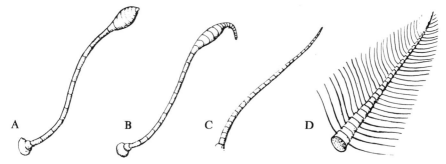

Figure 2. Antenae of Lepidoptera. A, B, butterflies; C, D, moths. A, *Euphy-dryas*; B, *Hesperia*; C, generalized filiform antenna of a moth; D, generalized pectinate antenna of a moth. (*Drawing courtesy of William H. Howe, from* Butterflies of North America, *page 1*)

at the tip. Other moths have wire-like antennae, and still others have small curves or elbows at the end of the antennae.

All Lepidoptera have two wings on each side of their body. The mechanism that joins the wings of most moths is called a *frenulum*, while butterfly wings function somewhat differently and are joined by a *humeral lobe*. We know that moths evolved before butterflies. The Australian Regent Skipper butterfly (*Euschemon rafflesia*) is unique in that it has a frenulum like a moth, which suggests that this butterfly might have been one of the earliest to evolve.

Because there isn't one single feature that absolutely distinguishes *all* butterflies from *all* moths, the classification of moths and butterflies merges on a continuum, and it can be subjective as to how a new species is classified. (See Chapter 4, Question 3: How is a species identified?)

Question 3: Why are they called butterflies?

Answer: The word butterfly comes from the Old English word *buterflage*, recorded in a glossary from about the year 700. It is a combination of the words "butter" and "fly," so it seems to be a description of a butter-colored flying thing.

Creamy white or pale yellow butterflies such as the Cabbage butterfly (*Pieris rapae*), the Clouded Yellow (*Colias crocea*), and the Brimstone (*Gonepteryx rhamni*) are common throughout Europe. These butterflies resemble the natural color of butter, before manufacturers started to add dyes to make the color deeper, so perhaps these butterflies were the inspiration for the name.

In ancient Greek, the word for butterfly is *psyche,* which also means *soul,* suggesting the mystical and spiritual aura that surrounds butterflies in many cultures. The origin of the Spanish word for butterfly, *mariposa,* or *Maria posa* (Posing Mary), refers to the praying hands of the Virgin Mary. In Aramaic (*parr*) the word for butterfly is derived from the word for flee. In Persian *parvani* is derived from the verb to fly (*parr*). In Sanskrit, *parna* is derived from the word for feather.

There is a word for butterfly in every language, and it is interesting to consider why the particular word has evolved in each language. The websites below have hundreds of additional examples of words that mean butterfly:

http://butterflywebsite.com/articles/saybut.htm

http://www.aworldforbutterflies.com/etymology.htm

The origin of the word moth may be from the Old English *maoa* or *moththe,* the Middle English *motthe,* or the Scandinavian *mott,* meaning maggot. It may also be from the root of the word midge, which until the 16th century was used mostly to describe larvae that devoured clothes.

Question 4: What is a skipper?

Answer: A skipper is a type of butterfly that belongs to the superfamily Hesperioidea. More than 3,500 species of skippers have been identified. A skipper typically has a hairy, stocky body, relatively small wings, a big head, hooked tips on the antennae, and a darting flight pattern. Many look very much alike and are difficult even for experts to classify on sight. Traditionally, skippers have been considered to be butterflies, but recently some experts believe they show more similarities to moths.

How Do You Say "Butterfly"?

Arabic (standard)	*farasha*
Balinese	*kupukupu*
Bengali	*prajapathi*
Chinese	*hu die*
Croatian	*leptir*
Dutch	*vlinder*
French	*papillon*
Gaelic	*feileacan*
German	*schmetterling*
Greek (modern)	*petalou'da*
Hebrew	*parpar*
Italian	*farfalla*
Japanese	*chou chou*
Mandarin	*hu-tieh*
Mandingo	*vrievran*
Senegalese	*lupe lupe*
Spanish	*mariposa*
Swahili	*kipepeo*
Zulu	*iveveshane*

More? See www.butterflywebsite.com/articles/saybut.htm

Question 5: How long does a butterfly live?

Answer: This question really refers to just one stage in the insect's life—the adult stage. (See Chapter 6, Question 14: What is the lifespan of a butterfly including all its stages?) Once the butterfly or moth has emerged from the pupa, the adult stage in most species lasts from a few days to a few weeks—just enough time to mate and lay eggs.

The tiny Spring Azure (*Celestrina ladon*) is one of the shortest-lived butterflies, lasting only a few days. Other species, such as the Mourning Cloak (*Nymphalis antiopa*), the Brimstone butterfly

(*Gonepteryx rhamni*), and the Compton Tortoiseshell (*Nymphalis vaualbum*), can live almost a year because they fly in the warm weather and then hibernate over the winter and emerge again in the spring to mate and lay eggs. The large Atlas moth (*Attacus atlas*) and related species have no mouthparts and so only live for about a week, using their body fat to sustain them since they are not able to eat or drink.

Diet also affects longevity. The Longwing butterflies (*Heliconius* spp.) live longer than those that only drink nectar because they have a more nutritious diet. Variations in lifespan also occur due to the hardships faced by an individual butterfly as well as because of general environmental and seasonal conditions. In conservatories and laboratories, butterflies tend to live longer than in the wild, due to the stable climate and the absence of predators, pesticides, and other environmental dangers. Normally short-lived Miami Blues (*Hemiargus thomasi bethunebakeri*) thrive for more than three weeks on melon-flavored Gatorade in a laboratory at the University of Florida. They even continued to lay eggs for much longer than they would normally.

Question 6: Does every butterfly of the same species look alike?

Answer: As with all living things, no two butterflies of the same species are exactly alike. Males and females of some species have strikingly different patterns and colors (this is called *sexual dimorphism*). The female Archduke butterfly (*Lexias dirtea*), for example, has an overall black and yellow pattern on its dorsal surface, while the male is black with a bright blue edge on its hind wing.

Some individuals vary due to their location, with color differences attributable to variations in diet, temperature, or day length during the caterpillar stage. If a caterpillar survives a shortage of food and pupates early, the butterfly that emerges may be healthy but smaller than normal. If the caterpillar has an abundance of food, the butterfly may be considerably larger than normal.

Generations of many species are lighter when they emerge in summer and darker when they emerge in colder weather. The fall or winter colors help camouflage the butterfly among dead leaves. The ability of an organism to make developmental changes in response to environmental conditions is called *phenotypic plasticity*. (See sidebar on Epigenetics: How Does the Environment Affect Butterflies?)

The female Eastern Tiger Swallowtail (*Papilio glaucus*) has two very different variations. The female normally resembles the male, with yellow and black stripes, but a black form of the female is more abundant in areas where the toxic Pipevine Swallowtail (*Battus philenor*) is more common. When a butterfly resembles a toxic species in order to deter predators, it is called a Batesian mimic, named after Henry Walter Bates, a mid-19th-century lepidopterist and collector who first wrote about mimicry. (See Chapter 7, Question 5: How do butterflies defend themselves?)

Once an adult butterfly has emerged, it cannot change its appearance—in contrast to many animals that have the ability to adapt their appearance to changing conditions. But individual butterflies do change in appearance as they age. They lose scales with every flap of their wings, and it is easy to see the difference between a bright, freshly emerged individual and one that is a week or two old with worn and faded wings. Some become quite ragged, showing tears and even bite marks on their wings. These marks may be used by researchers to identify individual butterflies.

Question 7: Which butterflies and moths are the largest in the world?

Answer: The largest butterfly in the world is the female Queen Alexandra's Birdwing (*Ornithoptera alexandrae*), which has a wingspan of more than 11 inches (28 centimeters). It is a toxic butterfly that lives in the rain forests of Papua New Guinea.

The Great Owlet or White Witch moth (*Thysania agrippina*), found from southern Brazil through Central America, has the largest wingspan, of up to 12 inches (30 centimeters), but the Atlas moth (*Attacus atlas*) from Southeast Asia has the largest wing surface area of all Lepidoptera and is considered the world's largest moth in overall size. Atlas females are larger than males, with a possible wingspan of up to 12 inches (30 centimeters). The Black Witch moth (*Ascalapha odorata*) and the Cecropia moth (*Hyalophora cecropia*) each have a wingspan of 5 to 6 inches, and are the largest moths in North America.

Question 8: Which butterflies and moths are the smallest in the world?

Answer: The Western Pygmy Blue (*Brephidium exilis*) is probably the smallest butterfly, with a wingspan that measures about 0.62 inches (0.25 centimeters).

There is a very large family of very small moths called Nepticulidae. The largest wingspan of any of the approximately 700 known species in the family is only 10 millimeters. The smallest member of this family that has been identified so far has a wingspan of just 2.5 millimeters (0.1 inch).

These tiny species are known as *micromoths,* and some of them are so small that it is virtually impossible to identify them with the naked eye. Sometimes it is necessary to dissect a specimen so that its unique body structures can be identified microscopically in order to determine its proper classification.

Question 9: How much does a butterfly weigh?

Answer: There is a thousand-fold difference between the weights of the lightest and heaviest butterflies. A very small butterfly weighs approximately 0.0001 ounce (0.003 grams), and a large specimen can weigh as much as 0.1 ounce (3 grams).

Question 10: What is a group of butterflies called?

Answer: There are several colorful words that aptly describe a group of butterflies. They are sometimes called a *kaleidoscope,*

Myths: Butterflies in Different Cultures

In Europe, butterflies were said to have been considered witches or fairies in disguise, stealing butter, cream, and milk. This could also be the source of the name butterfly.

Butterflies generally symbolize freedom, lightness and detachment, anxiety, or good luck.

In Greek mythology, Psyche, sweetheart of Eros, is often represented with butterfly wings.

Plinius, a Roman senator, is said to have thought that caterpillars developed when dew drops fell on a tree's leaves in spring.

In Hindu mythology, as Brahma observed metamorphosis he became filled with deep calm and was convinced that perfection could be achieved through rebirth.

The Taoist philosopher Chuang-tse was nicknamed the "butterfly philosopher" because he dreamed he was a butterfly and enjoyed flying around and sucking nectar.

In the Middle Ages big swarms of butterflies were sometimes seen as bad omens predicting wars and epidemics.

The belief that butterflies developed from the tears of the Virgin Mary comes from Romania.

In many countries pictures of butterflies can be seen on tombstones.

Some ancient Greeks considered butterflies to be the souls or spirits of dead persons.

The Slavs open a door or a window so that the dead person's soul, in the form of a butterfly, can leave the body.

In Finland some people believe the butterfly soul of a dreaming person flutters peacefully above the bed.

Goethe, the German poet, called butterflies "products of air and light."

In Korean symbolism, the butterfly represents happiness and conjugal bliss.

The Germans say a person in love has "butterflies in the belly."

which refers to a toy containing small, brightly colored tumbling objects that are reflected into symmetrical patterns by a set of mirrors. They are also called a *swarm,* a term usually applied to fish and birds as well as insects, describing a group of animals of similar size which are moving in the same direction. A group of butterflies is also called a *rabble,* which usually describes a mob or a disorderly crowd.

A group of Monarch butterflies (*Danaus plexippus*) is called a *congregation,* a rather orderly image. A group of caterpillars is called an *army.*

Question 11: Are there special words that describe people who love or hate butterflies and moths?

Answer: People who fear, are disgusted by, or generally dislike butterflies and moths are called *Lepidopterophobes* or *Mottephobes.* Mottephobia is defined as a persistent, abnormal, and unwarranted fear of moths, despite the understanding by the phobic individual and reassurance by others that there is no danger. Butterfly lovers are often simply called *Lepidopterists* or *Lepidopterophiles.*

TWO

Butterfly Bodies

Question 1: Does a butterfly have bones?

Answer: Butterflies and caterpillars do not have bones or a skeleton inside their body like we have. They have a fairly hard covering on the outside of the body called an *exoskeleton*. Lobsters and shrimp are familiar examples of other animals that have an exoskeleton.

The outer covering is made up of many layers of chitin and it hardens somewhat as it is exposed to the air when the butterfly emerges. It acts as a shock absorber, and it also makes the butterfly tough to chew on, helps keep out parasites and diseases, and keeps water inside the body so the insect does not dry out. The exoskeleton also lines the gut and tracheae (breathing system), thus also protecting these sensitive areas from damage and disease.

Question 2: How does a butterfly breathe?

Answer: One reason that insects are small is because their breathing system is inefficient and could not support a large animal. A butterfly breathes through nine pairs of pores or holes (*spiracles*) on the sides of the body. The largest spiracles are located on the thorax, where muscles from the legs and wings require more oxygen. These holes connect to a network of long air tubes (*tracheae*), and body movements pump the air through the tubes. Caterpillars and even pupae breathe in a similar way, through holes in their sides. The tracheae bring oxygen to the cells, and carbon dioxide simply diffuses straight out of the tissues through the exoskeleton.

Unlike many slow-crawling insects, butterflies need a lot of oxygen. A butterfly at rest requires three times as much oxygen per gram of body weight as a human. When flying, it needs 160 times as much.

Question 3: Does a butterfly have a heart?

Answer: A butterfly does not have a closed circulatory system made up of a central heart, veins, and arteries. It has a long, pulsating tubular heart along the top of its body that extends from its head to the end of its abdomen.

The internal organs are bathed in a fluid called *hemolymph* that circulates throughout the body cavity. Unlike our blood, hemolymph does not have to carry oxygen because air reaches the butterfly's organs directly through holes in its body. (See Chapter 2, Question 2: How does a butterfly breathe?) But hemolymph does act like blood in other ways, because it transports nutrients, wastes, hormones, and *hemocytes,* which are free-floating cells that play a role in the immune system.

Question 4: Do butterflies bleed?

Answer: Insects do have body fluid, called *hemolymph,* but they do not have red blood like ours. Our blood is red because it contains a special protein that carries oxygen. When butterflies emerge from their chrysalides, they release stored-up waste products that are often in a reddish fluid called *meconium.* This is sometimes mistaken for blood.

When the body of a caterpillar or butterfly is injured by a predator or bitten by another caterpillar because of crowded conditions or a lack of sufficient food, there may be some loss of body fluid at the point of injury.

Question 5: Do butterflies have good eyesight?

Answer: Butterflies have large compound eyes composed of hundreds of tiny units called *ommatidia*. They can see many colors as well as iridescent and ultraviolet patterns. They can easily detect motion, but they cannot focus their eyes and do not see details from a distance. Within short distances, they seem to see quite well. They generally avoid obstacles and are guided to the center of nectar flowers by ultraviolet patterns on the petals.

"Eyes," or groups of light-sensitive cells (*photoreceptors*), are found in unusual locations on many insects. Butterflies are no exception, as some males and females have photoreceptor cells on their genitals. This type of cell by itself can only simply sense the presence or absence of light, and their purpose in this location is not fully understood. (See Chapter 5, Question 5: How do butterflies mate?)

Question 6: Do butterflies have ears?

Answer: Some moths and butterflies have auditory organs, but they do not seem to hear very well. Most moths are nocturnal, which means they sleep during the day and fly at night. They are vulnerable to predatory bats, which locate the moths at night with ultrasonic hearing. Yack and Fullard studied moths that have "ears" that consist of a tympanal membrane like an eardrum, backed by an air-filled space and connected to a *chordotonal* organ with thin, elastic strands of connective tissue. The ears are usually located on the leading edge of the forewing, and they move with the wing during flight, enabling the moth to hear the echolocation calls of approaching bats and take evasive action. Other families of moths and butterflies have similar auditory structures on the thorax, abdomen, or at the base of the forewing and hindwing. Only some of these structures have been carefully studied and diagramed.

In some species the "ears" are located within the mouthparts. Some Sphingidae have fluid filled sacs alongside the proboscis (in a segment of the *labial palpus*) that has a limited ability to per-

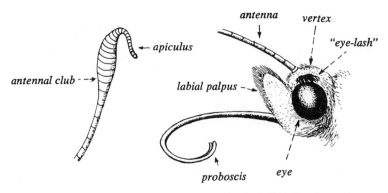

Figure 3. Head of a hesperiid butterfly (*Thorybes bathyllus*) showing major structures. Antennal club is to the left. (*Drawing courtesy of William H. Howe, from* Butterflies of North America, *page 12*)

ceive sound. Some larvae have sensory hairs (*sensilla*) that respond to airborne vibrations by triggering defensive movements.

Studies by Yack on male Cracker butterflies (*Hamadryas* spp.) found they also have hearing organs. Males make a cracking or clicking sound with their wing veins as part of their territorial displays, and other males can hear the sounds with an organ (Vogel's organ) located at the base of the forewing veins. This organ is similar to the structure of moth hearing organs described in the first paragraph.

Question 7: Do butterflies have a sense of smell?

Answer: They have scent receptors on their antennae, which are the primary olfactory organs. Many also have tiny, hair-like bristles called *setae* all over the body that are sensitive to smell, touch, and sometimes even to sound.

Question 8: How many legs does a butterfly have?

Answer: A caterpillar has eight pairs of legs. The forward three legs are attached to the thorax and become the adult butterfly's legs during metamorphosis. Behind these legs are five pairs of *prolegs*, which are attached to the abdomen. On the tip of each proleg is a ring of tiny, hook-like structures called *crochets*. These facilitate the caterpillar's movement, allowing it to easily cling to plant material, but they are discarded when the butterfly develops.

All butterflies, like all insects, have six legs, but you may see a butterfly that seems to have only four legs. In the family of "brush-footed" butterflies (Nymphalidae), the front legs are short and hairy and held close to the body. They are not used for walking and they may be very difficult to see.

Butterfly legs are made up of five jointed sections (*coax, trochanter, femur, tibia,* and *tarsus*). The clawed tarsus at the end of the leg serves as the foot, and it contains several segments itself, but usually not more than five. There are chemoreceptors (similar to taste buds) on the tarsi. When the tarsi come in contact with sweet liquids, the proboscis is stimulated to uncoil and sip the liquid.

Females have other chemoreceptors on their front legs, which they use to identify the proper host plant for their eggs. A female will drum her front legs repeatedly against a leaf. This punctures the leaf and releases a bit of the plant's aroma or juices. The special chemoreceptors on her front feet react to these substances and tell her whether the plant is a good place for her to lay her eggs. If it is suitable, the leaves will provide ready food when the baby caterpillars emerge from their eggs. (See Chapter 5, Question 9: Where do butterflies lay their eggs?)

Figure 4.   Leg of *Papilio polyxenes* showing major structures. Inset shows enlarged view of end of tarsus. (*Drawing courtesy of William H. Howe, from* Butterflies of North America, *page 21*)

Question 9: How many wings does a butterfly have?

Answer: A butterfly has four triangular wings: two forewings and two hindwings. They are attached to the thorax. Strong muscles in the thorax move the wings up and down in a figure-eight pattern during flight. Each region of each wing has a special name (e.g., *base, apex, tornus, costal margin, discal*, etc.), so that the wing markings and other details can be described in words without an illustration.

Question 10: What are butterfly wings made of?

Answer: Butterfly and moth wings are actually two transparent or brownish chitinous membranes. They are supported by tubular veins, and covered by thousands of colorful scales that overlap like shingles on a roof. The front and back of the wings often have different patterns and colors. Some species have clear patches on the wings with no scales, and there is a family of Clearwing butterflies (Ithomiinae) that have virtually no scales at all and delicate, transparent wings.

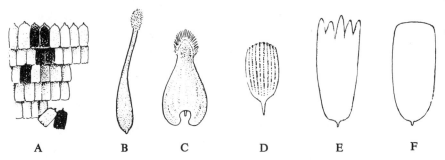

Figure 5. Wing scales. A, shingle-like arrangement of wing scales, *Papilio glaucus*; B, androconial scale, *Cercyonis pegala*; C, androconial scale, *Pieris rapae*; D, androconial scale, *Everes comyntas*; E, wing scale with toothed distal end; F, wing scale with entire distal end. (*Drawing courtesy of William H. Howe, from* Butterflies of North America, *page 18*)

Question 11: What makes the wings colorful?

Answer: The scales give color to the wings. Each scale is actually a hollow, flattened outgrowth of a single cell, and each scale is one color. Some scales are pigmented like paint colors, the color derived from chemical compounds called *flavonoids* that the larvae ingest from food plants and pass along to the butterfly. Pigment may also come from chemicals produced in the animal's waste products. Some scales have special surfaces that refract light, creating an iridescent effect, like the colors in a rainbow. They also absorb and reflect ultraviolet light, and many butterflies have ultraviolet patterns on their wings that are invisible to humans (because we cannot see ultraviolet light) but are attractive to likely mates (see Chapter 5, Question 3: How does a butterfly attract a mate?). If you touch a butterfly's wing, some scales will come off on your hand like a fine dust.

Question 12: What happens if a wing gets torn or damaged?

Answer: Once the butterfly has emerged as an adult, it totally stops growing and it does not have the ability to repair itself or heal its wounds, other than to stop the loss of body fluid due to a minor injury. We can heal injuries because our blood circulates to all the cells and makes the repairs. Butterflies have hemolymph instead of blood, it does not circulate, and it is a simple fluid without blood's complex healing properties. (See Chapter 2, Question 3: Does a butterfly have a heart?)

You often see older butterflies with damaged wings, but they usually can still fly quite well, even with pieces missing from the wing. In some species, tails at the back of the wing or eye-like spots on the wing actually draw predators' attacks toward the wing and away from the more vulnerable body. (See Chapter 7, Question 5: How do butterflies defend themselves?)

Question 13: Is it true that a butterfly will no longer be able to fly if you touch its wings?

Answer: Although butterflies are fragile, if you hold a butterfly properly you probably will not interfere with its ability to fly, although you will certainly cause it to lose some scales. The ability of the scales to come off easily is an advantage in getting free from sticky places, such as spiderwebs.

If it is necessary to hold a butterfly, you should hold it very gently with both wings together for as short a time as possible. If you are not careful it may try to struggle and injure a wing. Sometimes in a laboratory or a conservatory, a butterfly will actually step onto your finger if you present it properly.

Question 14: Why are butterflies called "cold-blooded"?

Answer: We warm-blooded (*endothermic*) creatures try to keep the inside of our body at a constant temperature by converting food we eat into energy in order to generate internal heat. When we are in a hot environment, we release heat into the air and cool ourselves by perspiring.

Cold-blooded (*ectothermic*) animals are hot when their environment is hot and cold when their environment is cold. They are much more active in warm weather and are very sluggish in the cold. Most butterflies are active between 60 and 108 degrees Fahrenheit (16 to 42 degrees Celsius). Many butterflies have to warm up to around 85 degrees Fahrenheit (28 to 30 degrees Celsius) in order to be able to fly. That is why you may see a butterfly sunbathing or *basking* in a sunny spot, trying to raise its body temperature.

Marsh Fritillary larvae (*Euphydryas aurinia*) spin a communal web at the base of their food plant, Devil's-bit Sacious (*Succisa pratensis*), which grows in a cool, marshy environment. The larvae are black and absorb solar radiation, allowing them to tolerate the relatively cold conditions. Being close together in the web is thought to raise their body temperature enough to enable them to digest their food. They leave in ones and twos to eat and to bask, returning to the web to digest their latest meal, to keep warm on cool days, and to overnight.

Moths that fly at night can generally fly at lower body temperatures. In the northeastern United States they have been observed to fly at less than 50 degrees Fahrenheit (10 degrees Celsius), and the long, fur-like scales on their bodies help trap whatever heat they manage to absorb or generate. Although they cannot warm themselves up at night by basking in the sun, they raise their body temperature by rapidly shivering their muscles, much as humans shiver in the cold in an instinctive effort to warm ourselves.

THREE

Butterfly Life

Question 1: Do butterflies bite?

Answer: Butterflies don't bite because they can't. Caterpillars munch on leaves and eat voraciously with their chewing mouthparts, and some of them do bite if they feel threatened. But once they become butterflies, they only have a long, curled proboscis, which is like a soft drinking straw—their jaws are gone.

There are, of course, always some exceptions. A few moths do not have any mouthparts (see Chapter 3, Question 3: What do butterflies eat?), and a very unusual group of very small, primitive micromoths (*Micropterigidae*) have chewing rather than sucking mouthparts, and they actually chew pollen in order to digest it. But they won't bite you!

Question 2: How do butterflies eat?

Answer: Almost every butterfly draws in nutrients by extending its curled proboscis, which unrolls when food is available, and sucking up nectar or other fluids. The proboscis works like a drinking straw, and when it is not in use, it is coiled up underneath the insect's head, between appendages called *palps*. Structurally, the proboscis is actually two filaments that stick together like Velcro.

Butterflies taste their food with special cells (*chemoreceptors*) on their feet (*tarsi*). A sweet substance detected by the chemoreceptors triggers the proboscis to unroll and take a sip. Other chemoreceptors on the proboscis confirm that the food is good to eat.

The length and nature of each species's proboscis are suited to its particular diet. The Giant Hawk moth (*Xanthopan morganii praedicta*) from Madagascar probably has the longest proboscis, at over 12 inches (30 centimeters), which is necessary because it drinks from deep, tube-like orchids. In most species, the proboscis is quite soft, only capable of sipping gently from a moist surface, but there are a few interesting exceptions.

Some fruit-eating moths, such as *Ophiusa tirhaca* and *Eudocima serpentifera,* have a sharp proboscis that can pierce citrus rind. The Vampire or Calpe moth (*Calyptra eustrigata*) from Southeast Asia has a barbed proboscis that can puncture thick-skinned fruits, and it can even pierce the skin of cattle and deer and feed on their blood.

Atlas moths (*Attacus atlas*) and related species in the family Saturniidae have no mouthparts and do not eat or drink anything during their short adult lives. They store up body fat in the larval stage, and they live for about a week until it has all been digested. (See Chapter 3, Question 5: Is it true that some butterflies and moths do not eat or drink?)

Question 3: What do butterflies eat?

Answer: Most butterflies get their energy from drinking flower nectar, which contains varying amounts of sugar. It is also rich in water, monosaccharides, and amino acids. Butterflies use nectar as an immediate energy source or store it as fat. Adequate nectar is usually essential for egg production (see Chapter 5, Question 10: How many eggs does a butterfly lay?). Each species has one or more plants that are its preferred nectar sources.

Some species feed exclusively on the juice from rotting fruit that has fallen to the ground. A few species, such as the Malachite (*Siproeta stelenes*) and the Rusty-tipped Page (*Siproeta epaphus*), feed on both flowers and fruit.

An interesting exception are the Longwing butterflies (*Heliconius* spp.), a group of toxic species from the American tropics.

Figure 6. Malachite butterfly (*Siproeta stelenes*) feeding on a banana.
(*Photograph by Hazel Davies*)

Pollen grains stick to the proboscis when the butterfly withdraws it from a flower after drinking nectar. The Longwing then secretes enzymes from the proboscis that mix with the pollen, and the pollen becomes a protein-rich liquid, which the animal absorbs. Digesting the amino acids in the pollen results in an unusually long adult life of six months or more.

Female Harvester butterflies (*Feniseca tarquinius*) lay their eggs among woolly aphids (*Neoprociphilus, Pemphigus, Prociphilus,* and *Schizoneura*). Woolly aphids, like other aphids, feed on plant juices and subsequently exude indigestible plant sugars as a sweet substance referred to as *honeydew,* which is consumed by the adult Harvesters.

Moths that feed on fungi have been found on sandy beaches in Saudi Arabia and on the shores of Lake Ontario in Canada, as well as in wooded areas in northeastern China and Finland. The Death's Head Hawk moth (*Acherontia atropos*), found in Europe, raids bee hives for honey.

Females of three species of tropical rain forest Ithomiine butterflies feed on bird droppings. The Purple Emperor (*Apatura iris*) and some butterflies in the *Vindula* genus feed on dung, and some *Vindulas* feed on dead animals and drink sweat. There are several skin-piercing, blood-sucking moths in Southeast Asia belonging to the *Calyptra* or *Calpe* genus. Commonly called the Vampire moth, the *Calyptra eustrigata* has a barbed proboscis and has been observed feeding on elephants, water buffaloes, and other large animals.

There are other Southeast Asian moths, such as *Lobocraspis griseifusa*, which are *lachryphagous*, meaning they feed on tears and other eye secretions of large, mild, plant-eating animals, and some have occasionally been observed drinking human tears. Butterflies were recently observed in Ecuador drinking the tears of Amazon River turtles. Although tear drinkers generally have soft-tipped proboscises, Hilgartner, a German entomologist, reported a fascinating observation of a moth, *Hemiceratoides hieroglyphica*, from a generally fruit-feeding or blood-feeding lineage of noctuids. This moth was seen attacking sleeping birds in Madagascar, sticking its sharply tipped proboscis into a bird's closed eye, presumably drinking the bird's tears.

Adult butterflies need minerals to complete their sperm and egg production, and salts and other minerals are passed from the male to the female in the sperm packet (*spermatophore*) (see Chapter 5, Question 5: How do butterflies mate?). It is necessary to replenish the minerals before the male can mate again, and male butterflies can be found taking in new supplies by feeding on a variety of salty and mineral-rich material (see Chapter 3, Question 6: What are butterflies doing when they gather on the ground? and Chapter 6, Question 4: What do caterpillars eat?).

Question 4: How do butterflies find their food?

Answer: Many flowers have ultraviolet patterns that are visible to butterflies. Since we cannot see ultraviolet light, these patterns are invisible to us. The patterns mark the center of the

What Do Butterflies Feed On?

Most butterflies feed on:
    nectar from flowers   rotting fruit

Some also feed on:
    muddy water   tree sap
    perspiration   animal dung
    mucus   bird droppings
    urine   decaying animal matter
    honey

A rare few manage to ingest:
    pollen that sticks to their   tears or eye secretions from
        proboscis   mammals, reptiles, or
    body fluids from aphids   birds
    blood from large animals

flower and guide the butterfly to the spot where it can quickly insert its proboscis into the nectar. This minimizes its exposure to passing predators when it is in a vulnerable feeding state.

Butterflies taste whatever they touch, using chemical receptors on the tips of their feet to identify a food source. If it tastes promising, the proboscis uncoils into the food. Moths that feed at night rely on sight and smell to find sources of food.

Question 5: Is it true that some butterflies and moths do not eat or drink?

Answer: Many moths have vestigial mouthparts and no digestive tracts so they are unable to eat or drink. These moths are in the family Saturniidae, along with approximately 2,500 species of lymantriid moths, including the Gypsy moth (*Lymantria dispar*), and some 300 species in the family Bombycidae.

The Luna moth (*Actias luna*), common in the eastern United States, and the Atlas moth (*Attacus atlas*) from Southeast Asia, are examples of large saturniid moths. They live for only about one week, during which time they mate and lay eggs, surviving on body fat accumulated in the larval stage.

Female Yucca Giant-Skipper butterflies (*Megathymus yuccae*), found mostly in the southern United States and northern Mexico, also do not feed as adults, but the males drink at muddy waterholes. (See Chapter 3, Question 6: What are butterflies doing when they gather on the ground?)

Question 6: What are butterflies doing when they gather on the ground?

Answer: Male butterflies require minerals not found in a diet of nectar or fruit juices, so they often drink water from puddles, stream banks, and other damp surfaces and retain the minerals dissolved in the water. If urine is present in the water, it will make the puddle especially attractive. This activity is called *puddling*.

Males pass minerals to females along with their sperm when they mate, and the minerals generally help sustain the females and their offspring, although in some species male-donated nutrients play little or no role in the female's potential reproductive capacity.

Question 7: Do moths really eat your clothes?

Answer: No adult moth is interested in eating your clothes. Moths typically only have a proboscis for sipping fluids and do not have the mouthparts necessary for biting or chewing. It is only the larvae of a very few species that are the guilty parties when it comes to eating fibers in clothing.

There are three types of "clothes" moths found in North America, members of the family Tineidae: the Webbing Clothes moth (*Tineola bisselliella*), the Casemaking Clothes moth (*Tinea pellionella*), and the Tapestry moth (*Trichophaga tapetzella*). The larvae of these species are very small—less than a centimeter long when at rest. They eat fibers such as wool, feathers, fur, hair, leather, lint, dust, and paper, and occasionally cotton, linen, silk, and synthetic fibers. The most damage is done to articles left undisturbed for a long time, and to fabrics that are stained with beverages, urine, oil from hair, or perspiration.

Question 8: How do butterflies excrete?

Answer: Like all animals, butterflies need to excrete bodily wastes in order to maintain a constant level of salts and water in the hemolymph as well as to get rid of toxic compounds that are produced when they digest their food. They excrete their wastes using a system of tubes called *Malpighian tubules,* which are long filaments that are found throughout the body. These tubes pick up waste materials from the hemolymph, and put the liquid waste (urine) into the hindgut (rectum) so that it can be expelled from the body. Although caterpillars have a fibrous diet and excrete solid wastes (frass), the wastes of a butterfly are usually only a few drops of clear liquid.

Question 9: Do butterflies sleep?

Answer: Butterflies do not have eyelids to close, so they never definitively look like they are asleep. Researchers have measured significant reductions in metabolism during periods of rest, and this is consistent with a sleep cycle. A period of sleep or rest is normal for all animals because it restores energy for activities such as eating and mating.

Sleep is tied to the 24-hour light/dark cycle (*circadian rhythms*) which regulates our biological clocks. Certain physiological and metabolic activities take place in the dark or during the resting phase, and others take place in the daylight or active phase. For example, many butterflies emerge from their pupae in the early morning. Some species are active during the day (*diurnal*), some during the night (*nocturnal*), and some are only active at dawn and dusk (*crepuscular*).

At night, many species of butterflies can be observed hanging motionless from leaves and twigs with their wings folded. Some species, like the Zebra Longwing (*Heliconius charitonius*) hang at night together in a large group, perhaps so they can rest or sleep more safely. Migrating Monarchs do this, too, although nonmigratory generations are usually solitary. And nocturnal moths have long periods of immobility during the day, which is their resting time.

Question 10: Do butterflies ever make noises?

Answer: Although most butterflies are silent, a few species do make startling noises in a variety of ways. Some moths communicate with sound: they make high-pitched squeaks and clicks in order to attract mates or to warn off predatory bats.

A butterfly may make a noise to discourage a competitor when it is searching for a mate, but noises are probably more often a way to frighten or at least to startle a predator so that the butterfly or moth can escape unharmed.

Male Cracker butterflies (*Hamadryas* spp.) from South America make a loud snap or series of clicks at both passing males and females while flying. Sometimes they will fly at other moving objects, such as leaves, other insects, or even passing humans. The clicking, audible to humans, may be produced by swollen veins in the forewings, which strike one another if the male forces the wings up as far as possible when he is flying, or perhaps by buckling and snapping wing veins.

The Death's Head Hawk moth (*Acherontia atropos*) blows air through its proboscis to make a loud squeak to discourage predators. The Tiger moths (Arctiidae) make noises by rubbing a wing against a leg, and the Corn Earworm (*Helicoverpa zea*) vibrates castanet-like structures on its wings.

In a startling discovery, Hay-Roe noticed that the blue and white Longwings (*Heliconius cydno*) in her laboratory were making faint clicking noises. She had to record the sounds to convince her skeptical colleagues that this well-studied species had this hitherto undetected attribute.

Question 11: Do butterflies communicate?

Answer: Yes, to a limited extent butterflies do communicate. Since the main focus of butterflies and moths is to reproduce, most communication is related to mating and is achieved using scent and movement. For example, in the case of nocturnal moths who are active at night and so have few visual cues, a female can indicate a readiness to mate by releasing a scent or pheromone to attract and guide males to her location (see Chapter 5, Question 3: How does a butterfly attract a mate?). Many species of butterflies and moths communicate sexual availability with pheromones, and some secrete defensive chemicals that repel competitors or predators.

In some species of butterflies, such as the Grayling (*Hipparchia semele*), the male will indicate his interest and try to persuade the female to mate by positioning himself in front of her and rubbing an aphrodisiac pheromone onto her antennae by catching the antennae between his forewings. Another form of communication occurs with elaborate courtship flights, the males fluttering above the females, or males and females landing together on a leaf and touching or stroking each other with their antennae or wings.

A female might say "no" to a male's advances with an upward spiraling flight in response to his approach. A male will tell other males to stay away from his potential mates by rapidly flying directly at an interloper and chasing him away, or by spraying the female, after they have mated, with an unpleasant scent that will repel other males.

Question 12: Can butterflies learn?

Answer: Butterflies operate by instinct, but, for example, they can learn to associate a color or a shape with food. Their color vision detects more wavelengths than humans can see, and research has shown that they can associate colors with rewards.

Researchers at Kyoto University found that Cabbage butterflies (Pieris rapae) could learn to forage increasingly quickly on a nectar source that was new to them, and they could remember a rewarding flower color for three days, even while they were learning that another flower color also offered nectar rewards. Pipevine Swallowtails (*Battus philenor*) learned a preference for yellow or magenta within ten visits to treat-laden flowers, in research conducted at the University of Arizona.

In another experiment, the second time butterflies tried to locate and sip the nectar from complex flowers they did it in half the time it took on the first try, apparently having learned how to insert the proboscis more efficiently. Learning is important because the environment can be unpredictable, and because each butterfly operates independently in finding food, an appropriate mate, and a suitable place to lay its eggs. The ease with which an animal can learn to adapt to changes in the environment determines the survival of its species.

Question 13: Do butterflies carry diseases?

Answer: There isn't a lot known about butterfly diseases, but we know that many viruses, bacteria, and fungi cause diseases in butterflies. One disease that was discovered in the late 1960s affects the milkweed butterflies, the Monarch (*Danaus plexippus*) and the Queen (*Danaus gilippus*). The protozoan parasite *Ophryocystis elektroscirrha* infects both wild and farm-raised populations, and can be debilitating. Individuals that are badly infected have difficulty emerging from the pupa and expanding their wings, or can be smaller and shorter lived than individuals without infection. The disease is often passed on when spores fall from the wings of infected females as they lay eggs. Newly hatched larvae are contaminated as they eat eggshells and milkweed leaves. There is no evidence of butterflies spreading any diseases to humans, but there is concern that the release of commercially raised butterflies at weddings and other events may spread diseases and epidemics to native butterfly populations (see Chapter 9, Question 17: Is it safe to release butterflies at weddings and other events?).

Question 14: What colors attract butterflies?

Answer: Experiments were conducted by scientists in Texas to determine if migrating Monarchs (*Danaus plexippus*) prefer flowers of a particular color, and the answer is *yes*. There is statistically significant evidence that the butterflies prefer orange flowers to yellow, red, and blue flowers. In one experiment the Monarchs were observed visiting orange flowers 37 times, yellow flowers 21 times, red flowers 5 times, and blue flowers 4 times.

Experiments with other species of butterflies have produced varying results, with some species appearing to prefer oranges, reds, and yellows, while others like whites, purples, and blues. If wishing to attract butterflies to your garden, plant flowers with a wide variety of colors to attract the widest variety of butterflies. Bright colors seem to be most attractive, but more importantly, large swaths of color will be most visible and make it easier for butterflies to locate your garden. (See Chapter 9, Question 4: How can I encourage butterflies to visit and breed in my garden?)

Question 15: Why are moths attracted to lights?

Answer: Some moths fly during the day (*diurnal*), and they do not have any particular reaction to light. It is thought that night-flying (*nocturnal*) moths use the light of the moon as a means to help them navigate. Because the moon is so far away, the moth perceives that the light rays come from the same general direction. By keeping the rays at a constant angle to its body, the moth is able to fly in a straight line.

Unfortunately, when a moth encounters an outdoor artificial light source, it tries to fly in a straight line, keeping the light rays to one side, but because the source is so close the rays are radiating out in all directions, and the moth keeps changing direction to compensate. It is possible that the moth's highly sensitive light sensors are overloaded by the intense light source. It is also possible that since flight is affected by heat, wing motion is reduced on the warmer, lighted side, causing the moth to turn toward the light.

The result of this confusing stimulation is that the moth flies around in circles and usually ends up bumping into the source of the light. Moth enthusiasts who want to study nocturnal specimens hang up a bed sheet with a bright light behind it and simply wait for the moths to come to the sheet as the daylight fades (see Chapter 9, Question 15: How can I see more moths?).

Question 16: How fast do butterflies fly?

Answer: Butterflies are very diverse in size and flight patterns, and their speed varies greatly. In general, they do not fly like airplanes. They have variable flight patterns, more like swimmers using various strokes.

Estimates of the flight speed of butterflies varies among different observers, but in general they find that toxic species tend to fly more slowly than nontoxic species. The fastest butterflies fly at over 30 miles (48 kilometers) per hour, and the slowest at about 5 miles (8 kilometers) per hour. Moths in the family Sphingidae (Sphinx moths and Hawk moths) are powerful and streamlined. Like the fastest butterflies, they have been observed flying at speeds of over 30 miles (48 kilometers) per hour, and when hovering to feed, have sometimes been mistaken for hummingbirds.

Monarch butterflies (*Danaus plexippus*) have been timed flying over 20 miles (32 kilometers) per hour. There may be other butterflies that are very fast flyers, but, unlike Monarchs, they have not been studied extensively.

Some Swallowtail butterflies beat their wings relatively slowly—only 4 or 5 times per second. Monarchs beat their wings from 5 to 12 times a second, still quite slow compared to other species who beat their wings up to 20 times per second.

Compared to the wings of most other insects, butterfly wings are quite large—which is why they beat their wings more slowly than other insects. In comparison, a housefly's wings beat at a rate of about 200 times per second, and mosquito wings beat about 1,000 times per second. What about hummingbirds? Their wings beat about 80 times per second during regular flight, and much faster during courting displays.

Question 17: Which butterfly can fly the longest distance?

Answer: The Guinness Book of World Records credits the world's longest butterfly migration to a Monarch (*Danus plexippus*) that had been tagged in Ontario, Canada, in September 1988 and recaptured in Austin, Texas in April 1989. Monarch expert Fred Urqhart was of the opinion that the butterfly had flown to Mexico and was on its way back north, having traveled an estimated 2,880 miles (4635 kilometers).

Question 18: How high do butterflies fly?

Answer: Bill Calvert of Journey North, a Monarch study group, reported that a glider pilot observed a migrating Monarch butterfly at a height of 11,000 feet (3,353 meters). There are other reports describing Monarchs passing overhead "at an altitude of one hundred meters or greater." Common Buckeye butterflies (*Junonia coenia*) have been recorded flying higher than 30 meters.

Information about altitude is mostly available about Monarch butterflies because their annual migration makes them excellent research subjects. There are lots of instructions available for Monarch migration watchers explaining how to observe and report flight patterns, but estimates of altitude are always rough approximations based on a range of methods, from simple (eyeball) to elaborate (radar).

The Insect Behavior Group at the University of Toronto, headed by David Gibo, has extensively studied Monarch flight patterns. They observed that migrating butterflies can often be seen bouncing around and flying erratically when turbulence in the air causes them to lose or gain altitude. Low-flying migrating Monarchs, flying within 3 meters of the ground, normally engage in straight flapping flight, sometimes alternating with straight gliding flight. When flying higher than 100 meters, straight gliding flight and straight soaring flight are common. Gliding butterflies always lose altitude, and soaring butterflies either maintain or gain altitude as they ride the air currents.

Question 19: Do all butterflies fly?

Answer: A group of researchers, headed by Angel Viloria, reported interesting results from a seven year project studying flightless butterflies in the Venezuelan Andes. Male Wood Nymphs (*Redonda bordoni*) in the Andes are competent flyers, but females of the species have markedly small, vestigial wings. Researchers found that when females were disturbed, they simply walked away, never attempting to fly. When one female was lifted up and dropped, she fluttered to the ground and then walked off. The condition in which wing size has been reduced to such an extent that flight is impossible is called *brachyptery*.

In the Andes species, the grass food plants that the larvae prefer are plentiful in the turf. The females just scatter their eggs while crawling around in the grass, and when the caterpillars emerge they are surrounded by food. The area where these flightless butterflies live is very cold and damp, and there frequently are very strong winds and fog, which make flying difficult. Thus flying is energetically costly, while flightlessness allows females to put more energy into egg production, strongly suggesting that environmental conditions are responsible for the female's evolution toward brachyptery.

Flightless Marion moths (*Pringleophaga marioni*), and a related species, *Pringleophaga kerguelensis,* live on the windy subantarctic Prince Edward Islands. In California's harsh Santa Maria dune system there is a species, the Oso Flaco moth (*Areniscythris brachypteris*), in which both male and female are flightless.

Female Fall Cankerworm moths (*Alsophilia pometaria*) have no wings and will crawl up a tree to seek safety. Some Geometrid moths, such as the female Tussock (*Orgyia* spp.), Inchworm (*Semiothisa bisignata*), Looper (*Erannis tiliaria*), and Spanworm (*Operophtera bruceata*), are also wingless.

A small number of female Schrankia cave moths (*Protanura hawaiiensis*) were observed in dark, high-altitude Hawaiian lava tubes and seemed to be consistently flightless, as is the Silkworm moth (*Bombyx mori*), which has been so domesticated that it has simply lost its ability to fly (see Chapter 6, Question 12: What is a silkworm?).

The female Vapourer moth (*Orgyia antiqua*) is wingless and stays by her cocoon. She attracts winged males with her scent, lays eggs on the cocoon, and dies there. The female Belted Beauty moth (*Lycia zonaria*) is also wingless, and her eggs survive with her body even after a bird eats her. The eggs pass out undamaged with the droppings of the bird.

Question 20: How do butterflies survive in cold climates if they need to be warm in order to fly?

Answer: Butterflies that live in cold climates have special characteristics that help them survive the cold. Nearly 100 species in the genus *Parnassius*, usually known as Apollos, live in the world's highest mountains, where it snows most nights (in the area of India, Pakistan, northern Afghanistan, and southwestern China.) These and other butterflies that live in cold climates have sorbitol or glycerol in their bodies, which keeps them from freezing. Producing body chemicals that prohibit the crystallization of vital body fluids is called *cryopreservation*.

Butterflies in the high Arctic and alpine regions tend to have hair-like scales on their bodies, which help them retain body heat, and they have dark areas at the base of their wings which absorb and maintain heat close to the body. The animals generally are small to medium sized, which minimizes their exposure to the cold. They fly low to the ground and can be grounded by the winds, and they select small depressions or sheltered spots in which to bask in whatever sun they can find. They fly only short distances and bask virtually constantly to maximize their thermal energy.

Many Arctic species freeze and refreeze during the larval stage. The time of the year when conditions exist so that it is even possible for them to fly at all is often restricted to about 30 days during July and August. But even during that warmer period, up to one-third of the days tend to be cloudy, so flight is not always possible even during the mildest conditions.

FOUR

Butterfly Background

Question 1: Why are scientific names used in this book?

Answer: Many animals and plants have different common names in different parts of the world. For example, *Nymphalis antiopa* is called the Mourning Cloak butterfly in the United States, but it is called the Camberwell Beauty in the United Kingdom. So that scientists everywhere can be sure they are referring to the same life-form, all animals and plants are grouped, or *classified*, so that each, after careful study, has its own unique scientific name.

Carl Linnaeus (1707–1778) is credited with establishing the idea of a hierarchical classification system, which is based upon observable characteristics, first published in his *Systema Naturae* in 1735. He introduced the concise *binomial* system, composed of a Latin name for the genus, followed by a "shorthand" name for the species, replacing the lengthy and nonstandard Latin names that were customary at the time. Although the binomial system was originated by the Bauhin brothers almost 200 years earlier, Linnaeus was the first to use it consistently, and he gets credit for popularizing it within the scientific community.

Linnaeus's hierarchical classification and binomial nomenclature, much modified, have remained the standard for over 200 years. The International Code of Zoological Nomenclature (www.iczn.org) contains the current classifications, and it is maintained and periodically revised by a commission dedicated

Carl Linneaus (1707–1778)

In the course of researching this book, coauthor Carol Butler had the privilege of being allowed to take photographs in the collection of the National Museum of Natural History in Leiden, The Netherlands. On plate 8 of the color insert you will find a photograph of a butterfly with a hand-written label indicating the date "1777"—the year it was identified, although it is known to have been collected earlier. It is a beautiful *Papilio ulysses*, found in an area ranging from the Moluccas to northeastern Australia and the Solomon Islands. The butterfly was named by Carl Linnaeus in 1758, and it is very likely that the person who collected it was influenced by Linnaeus and may even have encountered him in Leiden.

Carl Linnaeus was a Swedish naturalist who popularized and refined a system of hierarchical classification and binomial nomenclature of plants and animals that is the basis of the system used today. The unwieldy names which were commonly used at the time, such as *Physalis annua ramosissima, ramis angulosis glabris, foliis dentato-serratis,* were replaced with binomials, composed of the generic name, followed by a specific descriptor. In the above example, the new name assigned by Linnaeus to this rainforest herb is *Physalis angulata.*

Linnaeus lived in Leiden for almost a year during his three-year stay in the Netherlands (1735–1738). He was extremely influential and was considered perhaps the most renowned botanist of his time. In 1735, a senator from Leiden paid for the publication in the Netherlands of the first edition of Linnaeus' *Systema Natura,* an 11-page work. By the time the 10th edition was published in 1758, it classified 4,400 species of animals (including our butterfly) and 7,700 species of plants. Linneaus was ennobled by the Swedish king in 1757.

The system of classification was based on observable, shared physical characteristics, and his classifications have been revised as new ways of detecting scientifically valid observable characteristics have been developed. Organisms are presently being classified using DNA sequencing, and their hierarchical organization takes into consideration their evolutionary relationships.

to that purpose. (See Chapter 4, Question 2: How are butterflies classified?)

So that you can be clear about the butterflies we mention in this book, we use the scientific name in addition to the common name. If you enter the scientific name of a butterfly or moth into an Internet search engine that has "images" as an option, you will almost always be rewarded with several photographs of the species.

Question 2: How are butterflies classified?

Answer: The science of naming things is called *taxonomy,* and the classification system used for animals and plants was popularized by the Swedish naturalist Carl Linnaeus in 1735. Linnaeus used Latin names for genera and species because Latin was already the universal language of science. Modern scientific classification has become more complicated, describing the following levels of organization (*taxa*): Kingdom, Phylum, Class, Order, Family, *Genus, species, subspecies.* Scientific names are used in this book in the typical manner, specifying the genus first and then the species.

Butterflies and moths belong to the Animal kingdom (Animalia) and to the Arthropoda phylum. Members of a phylum may look quite different from one another to the casual observer, and so we find that in addition to butterflies, the Arthropoda phylum also includes crustaceans (e.g., shrimp), arachnids (spiders), and centipede-like insects. *Hexapods* are members of the subphylum of arthropods having six legs, which includes insects.

Butterflies are insects, members of the class Insecta. Insects are the only arthropods with wings. It is estimated that up to 70 percent of all organisms in the world are insects, making it by far the largest, most varied class in the Animal kingdom. Approximately 4,000 species of mammals have been identified, but, in comparison, over 750,000 species of insects have been named and described, and it is estimated that there are 5 million living species of insects. Although many insects play an important role in the ecosystem, they are almost universally re-garded as pests. Because butterflies are beautiful and graceful and do not bite, they are looked upon more favorably than other insects, even though their larvae can be agricultural pests.

Members of the class Insecta all have six legs and a three-part body consisting of a head, a middle section called a *thorax,* and a rear section called an *abdomen.* They are invertebrates, which means they do not have a spinal column or an internal skeleton like ours. Instead, they have an external skeleton (*exoskeleton*) that supports and protects the body (think of the outer skin of a shrimp, which is its exoskeleton).

Butterflies and moths belong to the order Lepidoptera. There are several families of macro and micro Lepidoptera, and within each family there are a number of *genera,* and within each genus are a number of species.

Here is an example of how this all works with regard to a specific butterfly. The Mourning Cloak, which you will see referred to in this book by its scientific name, *Nymphalis antiopa,* is a member of all the following taxa:

Kingdom	Animalia
Phylum	Arthropoda
Class	Uniramia (Insecta)
Order	Lepidoptera
Family	Nymphalidae
Genus	*Nymphalis*
Species	*antiopa*

Question 3: How is a species identified?

Answer: *Species* is a biological term meaning "type" or "kind." A simple definition of a species is that it is a group of animals or plants that look like each other. But a caterpillar looks nothing like the moth into which it metamorphoses, although they are the same species. Species that appear similar are not necessarily closely related, and closely related species may not resemble each other. In some butterfly species, male and female individuals look totally different (see Chapter 5, Question 1: How can you tell the difference between a male and a female butterfly?). Larvae can be useful in identifying a species, as can the larvae's choice of food. Detailed observations of physical and behavioral characteristics of all stages yield valuable information, which facilitates identification.

As Darwin was among the first to observe, species evolve over time as they adapt to environmental influences, such as diet and climate. This means that the appearance of butterflies belonging to the same species may vary within a local population, and the same species frequently looks quite different when found in different parts of the world. The traditional biological view has been that a new species diverges from its parent species when a population becomes isolated geographically or due to genetic or behavioral changes. The increasingly powerful focus on DNA mapping has established that gene flow between species is much more common than had been previously observed, and a growing number of hybrid animal and plant species are being identified. Cross-species mating (*hybridization*) does occur among butterflies, and its frequency is currently being studied. (See Chapter 5, Question 7: Do butterflies only mate with their own species?)

Reproductive capacity, meaning the ability to breed leading to healthy offspring, is a common criterion for defining a species. There was a "lock and key" hypothesis, which has been challenged as overly simplistic, which suggested that to avoid the production of maladapted offspring, females evolved specialized genital organs into which only the "keys" of males of the same species would fit.

In comparison with the relatively limited differences in external physical characteristics, moth and butterfly genitals, especially male genitalia, are so diverse and complex that they alone can be used to identify some species. Many authors of scholarly articles *only* illustrate the genitals when they are describing and classifying a new species. (See Chapter 1, Question 8: Which butterflies and moths are the smallest in the world?)

Eberhard compared male insect genitalia to a Rube Goldberg machine because of its fantastic complexity. Schilthuizen reviews the speculation that abounds as to the evolutionary significance of the organ's elaborate design, provoked, for example, by the observation that some isolated island species have evolved extravagant sexual organs although there is no chance of hybridization within the confines of the island habitat. (See Chapter 5, Question 4: How does a butterfly select a mate?)

DNA analysis provides a genetic profile of how an organism looks and functions, making it possible to define a species as a particularly stable combination of genes. There is presently a collaborative international effort called the Tree of Life Web Project (www.tolweb.org), which has the goal of analyzing and making available the DNA blueprints of all the organisms on Earth, essentially assigning a bar code to every plant and animal species. One result of this work is that from time to time, as more is discovered about a species, it has to be reclassified and moved to a different location on the evolutionary tree, which more accurately places it near related species.

Question 4: How many families of butterflies are there?

Answer: Butterflies are broadly classified into families according to their general shape and appearance. The five most universally recognized families of true butterflies are:

Papilionidae	the Swallowtails and Birdwings
Pieridae	the Whites and Yellows
Lycaenidae	the Blues and Coppers, also called the Gossamer-Winged
Riodinidae	the Metalmarks
Nymphalidae	the Brush-Footed

Experts differ as to which families should be included in the basic classification. Some other families that are considered impor-

Tree of Life Project

The Tree of Life Web Project is a collaborative effort of over 3,000 biologists from around the world. Its goal is to map the genome of every organism on earth and to establish their relationships to one another. Molecular analysis of over 1 million species has already been completed.

In combination with existing fossil data, a comprehensive species-level study of existing mammals has recently been published, and a similar analysis for birds was published previously. Eventually, every organism will essentially be assigned a bar code, creating a rich basis for more accurate future research and exploration.

Lepidoptera are the largest category of animals, and the "twig" of the project that is doing the molecular analysis of all Lepidoptera is hard at work and has already reported some new findings that have changed our understanding of how particular species have evolved.

On more than 4,000 web pages, the project provides information about the diversity of organisms on Earth, their characteristics, and their evolutionary history.

www.tolweb.org/tree is the home site for the project.

www.leptree.net is the twig of the tree project concerned with Lepidoptera.

www.lepbarcoding.org is the site of the All Leps Barcode of Life campaign.

tant by different entomologists are: Hesperiidae (the Skippers), Hedylidae (a very small neotropical family with 35 identified species), Libytheidae (the Snout butterflies), and many others.

There are huge numbers of moths of all different types, and their classification is very complex and is currently arranged into approximately 70 families. We will not describe that classification here, but we do discuss interesting moth species in every chapter.

Speciation: How Do New Species Arise?

The earth's rich biodiversity demonstrates that new species arise through evolution, but exactly how this occurs is a question that has intrigued scientists since well before Darwin. With the availability of molecular analyses that make it possible to precisely map the genome of an individual organism, what it means to "observe" the differences among species has taken on a new meaning. New discoveries indicating that mutations occur far more often than had been understood have challenged existing theories emphasizing that gradual adaptation is the path to genetic changes. Nevertheless, there are basic ways in which speciation occurs that are well documented.

Speciation commonly occurs when one species splits into two, in some cases because of geographic isolation resulting from changes in the habitat. A population is divided, and the groups are reproductively isolated. They stop breeding with one another and grow apart genetically in response to differing environmental stressors. For example, if a mountain range or a river divides a population of animals, perhaps for hundreds of generations, they may find that if they meet again they are no longer able to crossbreed. Even if some mating does occur, infertile offspring are typically produced, or the offspring may be fertile but less fit than their ancestors. This result confirms that a new species has evolved.

Although it is a less common occurrence, there are situations in which a new species results from crossbreeding of two distinct species. Usually when individuals from different species mate, speciation does not occur because their offspring die, are infertile, or are less fit. Fully sexual (*homoploid*) hybrid species are still considered to be quite rare in the Animal kingdom, although recent molecular analyses indicate that healthy hybrid butterflies are more common than had been expected. But even if cross species mating produces normal, sexual offspring, speciation does not typically occur because

(*continued*)

Speciation: How Do New Species Arise? (*continued*)

the offspring will mate with individuals from their parent species and the populations will merge.

As a general rule, sexual creatures prefer mates with familiar features and avoid mates with unusual or unfamiliar attributes. A research team studying *Agrodiaetus* butterflies in Asia headed by Lukhtanov found that if closely related species of that genus are geographically separate, they tend to look quite similar. But if closely related species share a habitat, they are likely to look strikingly different. This difference discourages inter-species mating and encourages genetic isolation and species divergence.

Another study by Gompert looked at two species of butterflies in the Sierra Nevada mountains. *Lycaeides melissa* lives on the eastern slope, and *Lycaeides idas* lives in the west. Researchers surveying the upper alpine slopes in the vicinity of Lake Tahoe found that a third species of *Lycaeides* had evolved in that area. The team used molecular genetics to show that the new species carries genes from both parental species. The scientists estimate that about 440,000 years ago the *L. melissa* and *L. idas* came into contact in the Sierras and their offspring eventually evolved into a genetically distinct species. The new species has a host plant on which it lays its eggs which is not the host plant for either parental species, further evidence that a habitat shift and speciation had occurred.

While *Lycaeides* produced a third alpine species in the Sierra Nevadas over the course of thousands of years, Mavarez reported that the evolutionary process has been recreated in just a few months. As reported in the June 15, 2006, edition of the journal *Nature,* a fully sexual hybrid species in the genus *Heliconius* resulting from crossbreeding two different parent species was created at the Smithsonian Tropical Research Institute in just three generations. There are more than 60 species of butterflies in this genus, and they exhibit a tremen-

(*continued*)

Speciation: How Do New Species Arise? (*continued*)

dous diversity of brightly colored wing patterns with red, orange, yellow, and/or white markings on a background of black. The color patterns serve as mating cues, and these butterflies are extremely choosey about finding mates with their own species-specific wing pattern. Butterflies with a yellow and red wing color pattern that is very similar to the wild species *Heliconius heurippa* were created through simple laboratory crosses of two other species. *Heliconius cydno,* which has a yellow wing stripe, was crossed with Heliconius melpomene, which has a red wing stripe. The genetic relationships among the three species were documented, and behavioral evidence confirmed that speciation had occurred because *H. heurippa* males only mated with females with both red and yellow stripes.

Question 5: How many species of butterflies are alive today?

Answer: Estimates of the number of living species of Lepidoptera range from 250,000 to 265,000. Estimates of the number of named and described species of butterflies range from 18,000 to 20,000, and more are identified each year. Because butterflies are large, colorful, and day-flying, they attract most of the attention, but they are a very small minority compared to the large number of moths that make up approximately 90 percent of the order.

The already identified species of butterflies are thought to account for most of the butterflies in the world. In contrast, some scientists estimate there may be a million or more species of moths, especially micromoths (very small moths), that have not yet been identified. (See Chapter 1, Question 8: Which butterflies and moths are the smallest in the world?)

Epigenetics: How Does the Environment Affect Butterflies?

In the wild, the larvae of the Pipevine Swallowtail (*Battus philenor*) are predominantly black at cooler temperatures and mostly red at temperatures greater than 86 degrees Fahrenheit (30 degrees Celsius). According to Nice and Fordyce, the red larvae are more tolerant of higher temperatures, and their growth rate does not slow down in the extreme heat, while the black larvae do not do as well when it gets really hot. This is an example of how genetically identical organisms reared under different environmental conditions can sometimes display diversity in physical characteristics and/or behavior.

Evolution has long been thought to occur primarily as a result of natural selection of the organism with the fittest set of genes, but natural selection does not act on genes—it acts on phenotypes. The term *phenotype* describes either the total physical appearance and constitution of an organism, or a specific trait, such as size, behavior, or coloring. This is a critical concept, because phenotypes are determined, in part, by the environment. When environmentally induced changes significantly increase the fitness of an organism, it is likely that natural selection will occur. Thus, the capacity of a phenotype to change (its *phenotypic plasticity*) plays an important role in evolution.

Traumatically shocking a butterfly with heat or cold during a critical period of its development can cause profound changes. First done experimentally in the 1890s by Merrifield, it was found that when the previously traumatized butterflies emerged from their pupae, their wing color patterns sometimes had changed to mimic patterns of related species that live in warmer or cooler climates. Wet season or dry season differences in wing patterns can also be induced by manipulating temperature during development.

In an 1896 experiment, the wing patterns of a heat-shocked Swiss subspecies of the Scarce Swallowtail (*Iphiclides*

(*continued*)

podalirius) appeared to resemble the normal form of the subspecies native to warm, sunny Sicily. In later experiments, heat shocking the central European form of the Old World Swallowtail (*Papilio machaon*) yielded some individuals that resembled both the Syrian and Turkish subtypes. Heat-shocked specimens of the central European subspecies of *Aglais urticae* produced wing patterns that resembled the Sardinian subspecies, while cold-shocked individuals of the central European variety developed the wing patterns of a subspecies from northern Scandinavia. Experiments with similar results have been done with the Mourning Cloak butterfly (*Nymphalis antiopa*), the Common Buckeye (*Junonia coenia*), the Western White (*Pontia occidentalis*), and several within the genus *Bicyclus*.

Other butterfly phenotypes that have adapted to experimental environmental manipulation are flight patterns (Speckled Wood butterfly caterpillars, *Pararge aegeria*), egg size (African Satyrid, *Bicyclus anynana*), pupae color (Nymphalidae, Papilionidae and Pieridae), body size (Great Eggfly, *Hypolimnas bolina*), and possibly oviposition.

Before molecular analysis was available, unexpected experimental results like these were considered embarrassing by some researchers. They were seen as examples of genetic defects that interfered with the research. By demonstrating with DNA testing that the genes remain stable despite the changes in appearance or behavior, the evolutionary relationships among phenotype, genotype, and the environment can be explored. One area of research is the study of the role of hormones in mediating the functioning of the genes. The ebb and flow of hormones regulates stages of development and the development of certain tissues, and research is focused on the role of hormones when the organism is subjected to extreme shifts in temperature.

Epigenetics is the field of biology that studies changes that can

(*continued*)

Epigenetics: How Does the Environment (*continued*)

be inherited but that occur "above and beyond the gene"—
that is, without any modification of the DNA sequence. One
of the ways that these changes are induced is in response to
environmental factors, such as temperature and diet. Some
of the mechanisms that facilitate these types of responses
are the increased activation of a single gene, the degree of
looseness with which the DNA "thread" is wrapped around its
protein "spool," and the variability of the location of the DNA
within the nucleus. These mechanisms enable the same gene
to function differently while remaining stable.

Question 6: When did butterflies and moths first appear on earth?

Answer: There is a good deal of variation among estimates
of when Lepidoptera first appeared. Current research estimates
that insects have been around for about 400 million years. The
fossil record of moths and butterflies is quite poor, and we know
very little about the earliest specimens and about how they may
have differed from modern species. The earliest Lepidoptera
fossil consists of three tiny wings of *Archaeolepis mane*, found on a
rock in Dorset, England, and estimated to be about 190 million
years old. This places them in the Early Jurassic era (206–180
million years ago), which means that primitive moths coexisted
with dinosaurs.

It is generally accepted that moths appeared before butterflies, but it is possible that butterflies and the larger moths
evolved from a common ancestor not yet identified. There is
agreement that by 40 million years ago all the main butterfly
families already existed.

Lepidoptera are strongly associated with the appearance (evolution) of flowers, since they are the largest group of organisms

that feed on plants, and the nectar from flowers is their primary food source in the adult stage. It has been said that "plants set the table for insects," because, even today, nearly half of all living insects depend on and interact with plants.

Recent research conducted in London and Shanghai by Cook using molecular sequence data (genetic analysis), as well as other studies, suggests a relationship between hexapods (the group of six-legged arthropods which includes insects) and crustaceans. This research hypothesizes that insects originated within the crustacean group, particularly a freshwater-dwelling group of crustaceans that includes water fleas and fairy shrimp, around 410 million years ago.

Question 7: Where are butterflies found?

Answer: Butterflies have been found in habitats ranging from lush rainforests to deserts and arctic tundra. They live on every continent except Antarctica, and they vary widely in their food preferences, their ability to tolerate temperature and moisture, and in their ability to travel to new areas. The tropics, especially the *neotropics* (the tropical regions of the Americas), contain the greatest diversity of species. (See Chapter 4, Question 8: Why are most butterflies found in the tropics?)

Most species of butterflies and moths are found in relatively small areas where there are specific conditions that meet their needs. Those that are more widespread, including species that are considered agricultural pests, have the ability as larvae to either eat a wide variety of food plants or have a preference for a food plant that is very widely available. At the opposite extreme, the Yellow Patch White (*Colotis halimede*) is adapted to the arid region in Africa south of the Sahara Desert. It often settles on bare earth.

Data collected about where butterflies are found are mostly the result of local butterfly counts, so the results usually apply only to a particular region or county. Counts are most valuable when done repeatedly over long periods of time to measure changes in a particular area. Butterfly distribution is constantly changing due to loss of habitat, global warming, and other factors. Regional distribution figures below have been gathered by many sources, and should only be used to generally compare the populations of different regions.

The Palearctic region includes Europe, northern Africa, the northern and central parts of the Arabian Peninsula, Asia north of the Himalaya foothills to the Arctic Ocean, and east to Korea and Japan. Transition areas occur along the southern borders, where elements of the Afrotropical and Oriental fauna fly together with Palearctic species. The climate in this region is mainly temperate, but it ranges from arctic to subtropical at its northern and southern extremes. Because the study of butterflies and moths began in Europe, the Lepidoptera of this region are better known than those in other parts of the world. A 2006 report in the Netherlands identifies 106 local butterflies, 55 of which are stable residents (the rest are migrants, vagrants, former residents, or irregular residents).

There is still limited knowledge about the fauna in the portion of the region that includes central Asia. Over 2,000 species of Lepidoptera have been identified in Israel alone, including 150 butterflies. The Arabian Peninsula is home to more than 150 species of butterfly. A 2002 count in Taiwan identified 377 species of butterfly, 56 of which are only found there (*endemic* to the region).

The Afrotropical region includes all of Africa south of the Sahara. This region has more than 2,500 identified species of butterflies. There are 43 species that are endemic to the Eastern Arc Mountains of Tanzania.

The Indo-Australian region includes Pakistan, India, Australia, and New Zealand. This region is mostly tropical and is one of the most populous areas for butterflies and moths. South India has reported 315 different species.

The Nearctic region ranges from Canada and Alaska to the southern border of the United States. It is mostly temperate, although there are extreme temperatures at the northern and

southern boundaries of the area. About 800 species of butterflies have been identified in this region, and many of these species are also found in the Palearctic region.

The Neotropical region begins at Mexico's northern border
and extends to the tip of South America. Approximately 2,000
butterfly species are found in Mexico. The tropical rain forests
of South America have the greatest diversity of species in the
world. Approximately 8,000 species of butterflies are estimated
to live in this region, although many have not yet been named
and described. Two-thirds of the species live in the Andes area,
which includes the countries of Venezuela, Colombia, Ecuador,
Peru, and Bolivia. Typical of the Caribbean islands, St. Kitts and
Nevis report 63 species in their small area.

Question 8: Why are most butterflies found in the tropics?

Answer: It has been said that one researcher estimated that
there are more types of insects in one tropical rain forest tree
than there are in the entire state of Vermont. The enormous
diversity of plant species provides a rich environment for a huge
diversity of plant-eating insects, making the tropics the most
biologically rich area in the world.

Compared with organisms in cool, dry habitats, the animals
in warm, wet ecosystems tend to have high metabolic rates and
most species produce several generations throughout the year.
Their fast-paced development increases the likelihood of rapid
evolutionary changes, genetic mutations, and the consequent
development of more species.

Just one of the many reasons that the warmth and moisture
in the tropics is particularly hospitable to butterflies and other
insects is that when they molt they must survive a period of
great vulnerability while the new exoskeleton hardens. (See
Chapter 6, Question 6: How does a caterpillar grow?) In a dry
climate, the soft body can become dehydrated without the hard

exoskeleton to hold in moisture, but in the humid tropics this risk is minimized. The largest insects occur in the tropics in part because this developmental danger is greatly reduced.

There are many tropical species that have yet to be named and identified, and as the loss of natural habitats in this area has accelerated, this region has become an important global priority for butterfly research and conservation.

FIVE

Butterfly Love

Question 1: How can you tell the difference between a male and a female butterfly?

Answer: Some butterflies are sexually *dimorphic,* which means it is easy to distinguish a male from a female because they look very different from one another. For example, in both nearly identical species of Archduke butterflies (*Lexias pardalis* and *Lexias dirtea*), the male is mostly black with a bright blue edge on the dorsal (upper) surface of its hind wings. The female Archduke has an overall black and yellow spotted pattern, and she looks totally different than the male except that the basic shape of their wings is the same.

Some males may be more colorful, some females may be larger, but in many species males and females look identical, or the external differences may be very subtle. Assigning gender in some species can be so difficult that it is necessary to rear a group of similar caterpillars in the laboratory in order to be able to identify the adult male and female butterflies by observing their mating and egg-laying behavior. Some very small species of moths can only be differentiated by actually dissecting each moth and inspecting its sex organs.

Question 2: Is it possible for a butterfly to be both male and female?

Answer: In rare circumstances, due to faulty cell division, a butterfly will be a mixture of male and female characteristics. If this occurs in a species where the male and female wing patterns are different from one another (*sexually dimorphic*), the wings on one side of the body may have the female coloration and the wings on the other side will have the coloration of the male. This is called a *bilateral gynandromorph*. In other individuals, the gender mixture may be more complex; this is known as a sexual mosaic. (See photograph on plate 8 of color insert.)

Question 3: How does a butterfly attract a mate?

Answer: Male butterflies usually search for females with whom to mate. Colorful ultraviolet wing patterns play a part in advertising the butterfly's gender. Patterns can change dramatically when seen from different angles, and in some species the males seem to try to position themselves in specific ways, presumably so that they look their most impressive as they try to attract available females. A male who is mistakenly approached

Gynandromorph

On plate 8 of color insert, you will find a photograph of a rare gynandromorph butterfly—a butterfly that is half female, half male. It is an Apollo (*Parnassius autocrator*), dated 1913, one of the prized possessions in the collection of Naturalis, the National Museum of Natural History at Leiden, The Netherlands.

The left half is female and the right half is male. The difference is clearly apparent in this dimorphic species because the coloring of the male and female are different from one another.

by another male usually flutters his wings vigorously to warn the approaching male to back off, although sometimes they interact competitively until one flies away.

Different species have different mating rituals. Some courting males engage in spectacular courtship dances, flying above or behind females, fluttering their wings rapidly. If the female is unwilling to mate with a particular male, perhaps because she has already mated, she may fly upward in a spiral path until the male gives up the chase. Some females will sit with the abdomen held in such a way that it is impossible for a male to approach them, and eventually the male will get discouraged and fly away.

Some males are territorial, patrolling a certain area, on the lookout for suitable females. They often dart threateningly at other male butterflies that approach their territory, and some even make sounds to scare away competitors (see Chapter.3, Question 10: Do butterflies ever make noises?).

Moths that fly at night, and therefore use less visual cues, often communicate with scent in order to locate a mate. Although in most species the male butterflies release pheromones (*sex hormones*) to attract or arouse a mate, it is often the female moths that sit and flutter their wings to move air over their scent glands to disperse the fragrant pheromones. Once near the female, male moths of some species also produce pheromones for a short period of time.

Some male pheromones may stimulate the female to mate, but others have different functions. Some pheromones inhibit the flight of the female and prevent resting females from flying away. These pheromones are composed of a dust that contains a flight inhibitor and a glue to stick the particles to the antennae of the female.

Each species releases a unique chemical blend, and *calling behavior* (the release of pheromones) varies from species to species. Pheromones are released at specific times of the day and with different rhythms or intensities, which probably increases the efficiency of the mating ritual because the search for a mate is limited and focused.

Some calling behavior is quite complex. For example, after

sunset, the males of some species of arctiid moths release their pheromones and attract other males, creating scent-emitting male groups. Females are attracted to the calling males, mating follows, and after an hour or two the male calling subsides and then the remaining virgin females emit pheromones, which attract available males. Jeremy Thomas observed that Brown Hairstreak males (*Thecla betulae*) congregate in an ash tree before mating. If their habitat does not contain a suitable "master tree," the mating process is disturbed, and indeed at the present time the species is endangered, at least in part because of habitat destruction.

In Lepidoptera, the pheromones are released from abdominal glands or by special scales on the wings. Pheromones released by male butterflies can only be detected over relatively short distances. The black spots (*androconial patches*) on the hind wings of male Monarch butterflies (*Danaus plexippus*) are actually hollow scales that are connected to scent glands which release pheromones. Male Green Birdwing butterflies (*Ornithoptera priamus*) have long scales that look like golden fringes on their hind wings for transferring scent during courtship. Male Paper Kites (*Idea leuconoe*) extend long yellow appendages (*hair pencils*) from the tip of the abdomen to diffuse their pheromones. Mitchell reported that when hair pencils were removed from male Angle Shade moths (*Phlogophora meticulosa*), mating was prevented. Some pheromones deter predators and competitors during courtship, and others are thought to induce the final stages of egg formation in the female.

Female Cecropia moths (*Hyalophora cecropia*) typically disperse a billionth of a gram per hour of their pheromone. Grondahl reports that the male moth can be attracted from 150 yards away, and there is an undocumented report on a website of a male moth that was marked and released, and traveled eight miles (17.6 kilometers) to reach a female.

A female Atlas moth (*Attacus atlas*) rarely strays from the area in the forest where she has emerged from her chrysalis, but she tries to perch in a spot where the air currents will best carry her pheromones. Males up to 3 miles (6.6 kilometers) downwind can detect her scent with the sensitive receptors on their long, feathery antennae. When she attracts a male, they quickly mate, she soon lays eggs, and both parents die in a few days (see Chapter 3, Question 3: What do butterflies eat?).

Question 4: How does a butterfly select a mate?

Answer: Sexual selection has been traditionally understood as male competition for sexual access to a passive female. The underlying assumption was that the competition takes place prior to mating, and that mating leads directly and more or less inevitably to fertilization of the female's eggs. Over the past 20 years, an impressive array of research in the area of *female choice* has been compiled that illuminates the magnitude of the female's role in determining the outcome of mating. Many species have been found to have mechanisms by which a female can accept a male for copulation but nevertheless reject him as a father, and butterflies are no exception. Some of the important sources in the literature about female choice are in Cordero, Eberhard, and Schilthuizen.

Female butterflies presumably select a mate based on some reproductive value they see in the particular individual male, perhaps his robust appearance indicated by bright colors, his energetic pursuit, or his optimal size. A female may exert control over the choice of a mate by raising her abdomen in a position that makes it impossible for a male to mate with her, or by otherwise avoiding his overtures. Because most postcopulatory competition among males for paternity is played out within the bodies of females, females will mate but may subsequently remove the spermatophore (the sperm packet). After mating, some males deposit a waxy substance in the female's opening (a *copulatory plug* or *sphragis*), attempting to make it impossible for her to mate again (see Chapter 5, Question 5: How do butterflies mate?). Females may reject or refuse the copulatory plug in order to make themselves available to a more desirable mate, but the presence of a sphragis confirms to observers that a butterfly has mated (see Chapter 5, Question 6: Are butterflies monogamous?).

Male butterflies perform courtship rituals that include visual displays, sounds, and scents (pheromones) that are employed both to bring the male within mating distance of the female and also to repel competing males. Males develop characteristics that make them attractive to females, and some develop attributes which do not seem to interest females, and whose purpose seems to be to enhance their ability to compete with other males. These attributes give them a reproductive advantage over other males, and this advantage can be inherited by their male offspring.

Male butterflies and moths have very elaborate genitalia, and some scientists speculate that the genitals, in addition to delivering sperm, may be an "internal courtship device" providing tactile stimulation to excite the female and encourage her to copulate (see Chapter 4, Question 3: How is a species identified?). When a competing male mates with a female who has already mated, sperm competition may take place within the female reproductive tract. The most recent consort may use physical or chemical means to remove or destroy the sperm previously deposited by another male. In other species, chemicals accompanying the sperm can act as an anti- aphrodisiac on the female, reducing her interest in subsequent mating. Some males deposit chemicals on the female externally to repel other males, discouraging them from finding the female sexually attractive.

Question 5: How do butterflies mate?

Answer: The male butterfly or moth has a pair of claspers at the end of his abdomen with which he grasps the female, and when they mate they join end to end and tend to remain in that position for several hours. If disturbed, one butterfly may fly away with the attached partner dangling below. Prolonged copulation in butterflies (and in other animals) delays or prevents subsequent mating with other males and allows time for the male's sperm to be used for fertilizing the female's eggs, thus giving a reproductive advantage to the male as long as he and the female remain joined.

The male passes a sperm packet through his penis into the

Figure 7.   Mating Paper Kite butterflies (*Idea leuconoe*). (*Photograph by Carol Butler*)

female's reproductive tract, and she fertilizes her eggs at a later time when she finds a suitable food plant on which to deposit the eggs. The sperm packet also contains minerals to nourish the eggs when they are fertilized, and it may contain chemicals to discourage subsequent mating. (See Chapter 5, Question 4: How does a butterfly select a mate?)

In some species, males emerge first from the pupa so they are ready to mate as soon as females are ready to emerge. Gilbert reports the Postman butterfly (*Heliconius erato*) patrols his area looking for a chrysalis containing a female that is about to emerge. He will sit on the chrysalis so that he can mate with the female while she is emerging, before she can even fly. After mating he deposits a chemical on the female that repels other potential suitors. Other males, after mating with a female, plug her opening with a waxy substance (a *copulatory plug* or *sphragis*) in an attempt to prevent her from mating again.

Male and female butterflies have light receptor cells in the area of their genitals (see Chapter 2, Question 5: Do butterflies have good eyesight?), and one hypothesis about the function of these cells is that if the fit when the butterflies try to mate is not perfect, light will be detected and the mating will be aborted.

Some species, for example the female Atlas moth (*Attacus atlas*) and the Finnish Bagworm moth (*Dahlica lichenella*), are capable of reproducing without mating (*parthenogenesis* or *virgin birth*). These females can duplicate an egg cell as if it has been fertilized, but the eggs they lay only yield female offspring.

Question 6: Are butterflies monogamous?

Answer: As in most species, male butterflies tend to mate more than once. Males lose salt and other nutrients in the sperm packet (*spermatophore*) when they mate, and in order to mate again they ideally need to replace these nutrients by drinking mineral-rich liquids (see Chapter 3, Question 6: What are butterflies doing when they gather on the ground?).

To ensure that their own sperm fertilizes the female, rather than the sperm of a prior or subsequent suitor, male butterflies use various strategies to eliminate other males' sperm (see Chapter 5, Question 4: How does a butterfly select a mate?).

Females that only mate once are *monandrous*. Female Regal Fritillaries (*Speyeria idalia*) observed in Kansas have a lifespan of up to eight weeks but mate only once shortly after they emerge from the pupa. In other species, females are *polyandrous*, which is to say they mate with more than one male. A polyandrous female may internally select the sperm from the healthiest male to fertilize her eggs, or she may have a genetically mixed batch of offspring.

Question 7: Do butterflies only mate with their own species?

Answer: Butterflies usually only mate with their own species, but mating across species does occur, although it is unclear in general how often it results in fertilization and healthy off-spring. A 2007 report in *BMC Evolutionary Biology* estimates that 26–29 percent of *Heliconius* butterflies hybridize in the wild.

The historical viewpoint has been that hybrids tend to be sterile or less healthy than homogenous individuals. In fact, there are many known butterfly hybrids that successfully mate with each other and with their parents' species. And in general, the increased use of molecular biology has revealed that hybridization is much more common among plants and animals than had been generally observed.

Factors that prevent crossbreeding are anatomical (the two species do not fit together), chronological (they may fly at different times), geographical (they fly in different locations), and behavioral (their typical courtship behavior only attracts suitable mates of the same species) (see Chapter 5, Question 3: How does a butterfly attract a mate? and Chapter 4, Question 3: How is a species identified?).

Question 8: What does a butterfly egg look like?

Answer: Butterfly and moth eggs are very small, usually 0.1 inch (3 millimeters) or less, and they vary greatly in size and shape. Their shape can be spherical, oval, or pod-shaped; some are clear and some are white, green, or yellow. Some have a smooth surface and some are ridged or have an otherwise irregular surface.

A butterfly egg has a hard, thin outer shell (*chorion*). The shell is lined with a layer of wax that helps it retain moisture. A yolk inside each egg provides food for the developing larva. Each egg has one or more tiny funnel-shaped openings at one end (*micropyles*), which allow sperm to enter the egg. While the egg is developing, air and water enter the egg through the micropyle.

When it is time to hatch, the larva chews a hole in the egg and crawls out. It usually eats the remainder of the egg case, which supplies nutrients as well as symbiotic bacteria that help the larva digest the cellulose in its diet.

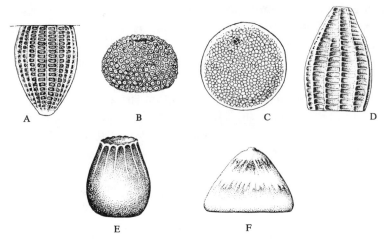

Figure 8. Eggs of different butterflies. A, *Danaus plexippus* (after Scudder);
B, *Eumaeus atala* (after Rawson); C, *Euptychia mitchellii* (after Hubbell);
D, *Anthocharis midea* (after Scudder); E, *Euphydryas phaeton* (after Scudder);
F, *Atrytone arogos* (after Howe). (*Drawing courtesy of William H. Howe, from*
Butterflies of North America, page 4)

Question 9: Where do butterflies lay their eggs?

Answer: Each species of butterfly is dependent on a fairly
specific habitat in order to reproduce successfully and to develop its eggs to maturity. A female will almost never lay her eggs
(*oviposit*) unless she finds a suitable food plant that will be readily available for her larvae when they emerge from their eggs
(see Chapter 9, Question 6: What is a host plant?). Most larvae
eat leaves, grasses, and other forms of vegetation from a specific
plant, and the adult female must somehow confirm that a plant
is a suitable host. In many species, the female not only chooses a
host plant, but also chooses the age of the plant, its density, and
a specific location on the plant so that the egg ideally releases
the larva near a fresh, edible bud.

Female Olivencia Tigerwings (*Forbestra olivencia*) have been
studied by Hill when they are searching for a place to lay their
eggs. They do an elaborate assessment of a potential host plant,

using all their senses. They hover around the leaves and stems, and descend slowly along the main stem almost to the ground and then ascend back to the leaves. They have been observed touching the leaf surface with the antennae and the abdomen. One female was reported to have taken ten minutes to make this evaluation.

Adult butterflies and moths usually have only a soft, straw-like proboscis, which is fine for sipping nectar but is not suitable for tasting the leaf of a host plant. There is ongoing research whose goal is to completely decipher the "tasting" behavior, but a general explanation is possible.

Plants are very active, and their dynamism plays an important role in the life of a butterfly. Plants give off bouquets of fragrances, which serve various functions. Some odors attract butterflies and moths, helping to direct them to the valuable nectar food source. Other odors repel insects that might harm the plant, and still others attract predators of insects that use the plant as a food source.

Plants that serve as hosts for butterfly eggs may have leaves coated with waxlike substances, which strive to mask the attractive odors that indicate that the plant is a suitable host. Adult females have groups of sensory hairs (*trichoid sensilla*) equipped with taste and touch receptors on their legs. They also have spine-like structures close to the sensilla, which may have a touch function. They drum their front legs against the leaf of a potential food plant, and when the drumming penetrates the protective layer, substances are released that are assessed (perhaps not quite "tasted" or "smelled" but somehow sensed) by the hair-like sensilla. If the sensory input is recognized as belonging to a suitable host plant, the female will oviposit one or more eggs.

Some plants have other defenses that discourage butterflies from laying eggs on them, thereby minimizing the number of caterpillars that will eat its leaves. The *Passiflora* or passion flower vine is the food plant for Longwing butterflies (*Heliconius* spp.). These plants produce small glands called *foliar nectaries* that look like Longwing butterfly eggs. The presence of the nectaries serves to discourage females from laying their eggs on a

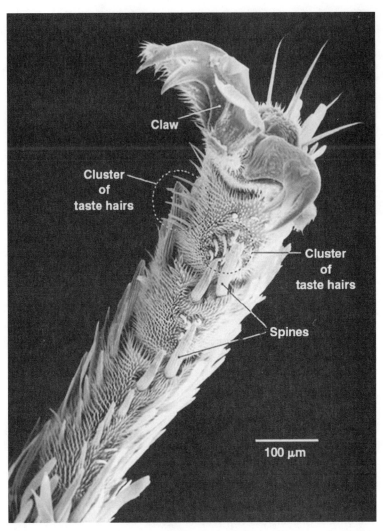

Claw

Cluster
of
taste hairs

Cluster
of
taste hairs

Spines

100 μm

Figure 9. Front leg of a butterfly showing sensory hairs and spines that enable her to identify a host plant by tapping on a leaf and assessing the volatile chemicals that are released. (*Used with permission of Dr. Erich Städler, Department of Environmental Sciences, Conservation Biology, University of Basel, Switzerland*)

particular leaf because it already appears to be occupied. The nectaries also attract ants that prey on Longwing eggs and caterpillars, further minimizing the potential damage to the plant caused by the voracious caterpillars.

Grass-feeding butterflies like the Ringlet (*Aphantopus hyperantus*), Marbled White (*Melanargia galathea*), and Woodland Brown (*Lopinga achine*) release their eggs as they fly low over the vegetation. Their larvae find food when they hatch from the eggs in the grassy turf.

Some root-boring moths lay their eggs on bark near the base of the plant, and the young larvae burrow downward in aboveground stems and into the root crowns in later instars. They pupate in the root, head upward, and emerge through a frass-covered track to above-ground level.

Question 10: How many eggs does a butterfly lay?

Answer: In some species the females lay one egg at a time, others lay eggs in small clusters, while others lay large numbers at a time. The Speckled Wood butterfly (*Pararge aegeria*) lays eggs one at a time. The Map butterfly (*Araschnia levana*) lays its eggs in columns.

The Greater Wax moth (*Galleria mellonella*) may lay over 1,000 tiny eggs (about 0.5 millimeters in diameter), usually in batches of about 100. Huge clutches of eggs, as many as 2,000 on a leaf, are probably the result of groups of females laying together (*gregarious oviposition*). These large clutches may have adaptive value in preventing the eggs from drying out and providing some protection against predators.

Many factors influence the average number of eggs that are laid by each species and by an individual butterfly. Some butterflies are *capital breeders,* using reserves accumulated during larval growth to develop their eggs, so an adequate supply of host plants is especially important for reproduction to occur. Other butterflies are *income breeders* who depend on adult foraging in order to acquire the necessary reserves to develop their eggs, so availability of nectar plants to nourish the adult butterflies is extremely important for these species to maintain their population by producing sufficient eggs.

Laboratory research on the Mormon Fritillary butterfly (*Speyeria mormonia*) sponsored by Stanford University showed that butterflies on a restricted diet laid fewer eggs. Instead of laying eggs, the adults used up the rich nutrients in the immature eggs in order to survive (*oocyte resorption*), indicating that egg maturation and longevity are affected by a scarcity of nectar.

Regal Fritillaries (*Speyeria idalia*) were observed by Kopper to lay a very large number of eggs in the Kansas tallgrass prarie. One female under observation laid more than 2,450 eggs over her 12-week adult life. Females of this species lay their eggs in late summer, when the food plants for their larvae (violets) are only withered leaves. The females lay eggs on the dry leaves and on any dried vegetation and even in the soil. In the laboratory, they were observed to lay eggs on netting and toweling. The larvae hatch a short time after the eggs are laid, and since there is no food available, they go into diapause until the food plants revive (see Chapter 7, Question 13: How do butterflies survive harsh weather?).

Temperature also plays a role in egg production. Fischer experimented with African Satyrid butterflies (*Bicyclus anynana*), rearing them at different temperatures in an otherwise uniform environment. Females reared at a lower temperature laid significantly larger and fewer eggs than their sisters kept at a higher temperature.

Laying large batches of eggs is nature's attempt to ensure that at least a few eggs will survive the attacks of the many predators for whom the eggs are an attractive food source. Most eggs are attached quite firmly to a leaf, a stem, or a flower with a fast-drying glue-like chemical that the female butterfly secretes along with the egg. Alpine lycaenid butterflies are an interesting exception because they attach their eggs quite loosely to the host plant. In their mountainous habitat, almost all of each year's plant growth is blown away by strong winds during the winter months, so it is adaptive that their eggs are not attached strongly and can easily fall off the plant to relative safety on the ground, where they await the new growth in the spring.

Question 11: How long does it take for the eggs to hatch?

Answer: The development of the larva inside the egg usually takes a few days to a week or two, depending on the species. However, host plant availability is a factor in the timing of the larvae's behavior, and some species of butterflies and moths spend the winter in the egg stage, waiting to emerge when milder weather signals that food plants are readily available (see Chapter 7, Question 13: How do butterflies survive harsh weather?).

SIX

Metamorphosis

Question 1: How does a caterpillar become a butterfly?

Answer: Butterflies and moths pass through three very distinct life stages after emerging from the egg: the larva or caterpillar stage, the pupa, and the adult butterfly or moth. The process of development during which this change occurs is called *metamorphosis* (see Chapter 6, Question 2: What is metamorphosis?). A complex process divides the developmental tasks between two highly specialized creatures; the caterpillar's task is to eat and grow, while the butterfly is focused on dispersing into new areas and reproducing. The pupal stage is considered a resting stage, where the animal does not move or eat, although great changes occur and there is some movement that occurs spontaneously or in reaction to touch (see Chapter 6, Question 8: What happens inside the pupa?).

Question 2: What is metamorphosis?

Answer: An apt definition of metamorphosis is "a transformation, as by magic or sorcery." The word metamorphosis comes from the Greek word *metamorphoun*, meaning to transform, a compound of the words *meta* "after or beyond" and *morphe* "form."

In biology, metamorphosis means a pronounced change in both the internal and external form (morphology) of an animal during normal development. These changes are triggered

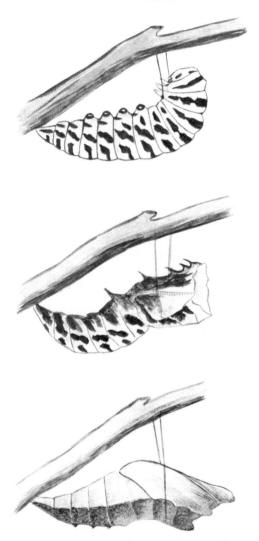

Figure 10. Three stages in the transformation from caterpillar to pupa.
The top illustration shows the caterpillar suspended from a branch by a silken
thread. The center illustration shows the soft chrysalis appearing from under
the old split skin of the caterpillar. The third illustration is the pupal stage,
with the caterpillar enclosed in the chrysalis. (*Drawing courtesy of William H.
Howe, from* Our Butterflies and Moths, *page 26*)

by a combination of cues from hormones, genetic mechanisms, the length of the day, the temperature of the air, the amount of moisture in the air, and other variables. The animal's habits or needs also often change profoundly because of the physical changes. In the example of the caterpillar-to-butterfly metamorphosis, the animal becomes able to fly, and it can no longer bite or chew so its diet must change completely.

Metamorphosis is not unique to butterflies; most insects undergo some type of metamorphosis. There is an entire group of *holometabolous* insects that undergo complete metamorphosis, including beetles, flies, ants, and bees. The change from a tadpole into a frog is an example of complete metamorphosis in an amphibian. An evolutionary advantage of metamorphosis, which might partly explain why there are so many living insect species, is that the immature and adult stages are often so completely different that they do not compete with each other for food, as would occur in a continuously developing animal. A silverfish is an example of a continuously developing animal. They hatch as a small version of their adult selves and just grow larger. They do not metamorphose, and the adults differ from the immatures only in size and in possession of fully developed reproductive organs.

About 12 percent of insects go through incomplete metamorphosis, including crickets and grasshoppers. Incomplete metamorphosis is characterized by the absence of a pupal or resting stage between the immature and adult stages. A female insect lays her eggs and the eggs hatch into *nymphs*. Nymphs look like small adults but usually do not have wings, and they eat the same food that the adult insect eats. Nymphs molt several times as they grow, and they stop molting when they reach their adult size. By this time, they have usually grown wings. In the case of other insects, such as mayflies, dragonflies, and damselflies, the immatures are known as naiads. Naiads live in the water and breathe through gills. After they have molted and reached adult size, they fly and breathe air, so their diet and habitat are totally changed.

Question 3: What is the difference between a caterpillar and a larva?

Answer: Larva (plural is *larvae*) is another word for the caterpillar stage of a butterfly or moth. A caterpillar or larva is an eating machine whose only functions are to eat and grow.

Caterpillars have three distinct body parts: a head and a two-part body consisting of a thorax and an abdomen. The head has a pair of very short antennae, mouthparts (upper lip, mandibles, and lower lip), and six pairs of very simple eyes, called *ocelli* or *stemmata*. Even with all these eyes, the caterpillar's vision is poor. There are sense organs in the mouth (*maxillary palps*), which help guide the caterpillar to its food.

Three pairs of jointed "true" legs are attached to the thorax, and five pairs of false legs, or *prolegs,* are attached to the abdomen. The prolegs have tiny hooks on them that facilitate movement and hold the larva onto its leaf or silk mat. They are discarded during metamorphosis. (See Chapter 2, Question 8: How many legs does a butterfly have?)

Question 4: What do caterpillars eat?

Answer: Almost all caterpillars are voracious eaters, growing quickly to as much as 1,000 times their birth weight. But not all caterpillars grow quite so dramatically. If they eat tough grasses

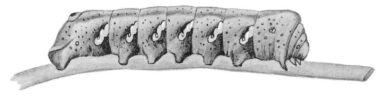

Figure 11. The slug-shaped caterpillar of the Achemon Sphinx (*Eumorpha achemon.* (*Drawing courtesy of William H. Howe, from* Our Butterflies and Moths, *page 140*)

and woody substances, chewing and digesting take more time and energy than if they eat leaves, and so they grow more slowly. The slow-growing Carpenter Worm (*Chilecomadia valdiviana*) is a caterpillar that feeds on wood. It tunnels into trees such as willow, oak, maple, and elm and can take three years or more to reach full size. The first meal for a caterpillar is usually its egg case. The case provides the symbiotic bacteria the caterpillar needs to digest the cellulose in its leafy diet as well as some nutrients that are essential for the insect's growth.

Although there are exceptions, like the Yucca Giant-Skipper (*Megathymus yuccae*), which eats the roots of the yucca plant, most caterpillars feed on the leaves of specific plants that are their host plants. Caterpillars can sometimes be found on plants that are not their hosts, but this usually means they are diseased, parasitized, or ready to pupate.

Larval Orchard Swallowtails (*Papilio aegeus*) energetically chew the leaves of orange, tangerine, and lemon trees. These noisy feeders can be located at night by their loud leaf-crunching. Vanilla, the only orchid that is a climbing vine, is the preferred food plant of a sub-Sahara African species (*Palla ussheri*), putting this species in the category of being an agricultural pest. Many other species are also considered agricultural pests because their food plants are valuable crops. (See Chapter 9, Question 2: Is it true that some butterflies and moths have a negative impact on the environment?)

Larval Eastern Tiger Swallowtails (*Papilio glaucus*) eat the leaves of a wide variety of host plants, including wild cherry, basswood, birch, cottonwood, mountain ash, and willow. Although it is not unknown for caterpillars to adopt new food plants, some species that are unable to find their preferred food plant will starve rather than switch to other plants that are present in their habitat. Those that cannot switch are called *specialists*, and those that will accept more than one host plant are called *polyphagous*.

Some caterpillars have unusual eating habits. Deyrup describes a primitive moth known as a horn feeder (*Ceratophaga vicinella*) whose larvae feed on the keratin shells of dead gopher tortoises. Although ants are among the leading invertebrate predators, many insect species interact with ants in a relationship that is mutually beneficial (*myrmecophily*). The Lycaenidae is a large family of butterflies that includes Blues, Coppers, and Hairstreaks. More than half of the species of lycaenid butterfly larvae live peacefully with ants, usually excreting sugary, carbohydrate-rich droplets from a gland on their backs that the ants consume. The ants, in return, patrol the surrounding plants and attack predators and parasitic wasps that approach the caterpillars. Some larvae release chemical signals that warn the ants of impending danger and some make warning sounds. Caterpillars of some other species eat the immature ants (eggs, grubs, and pupae). Five such species live on the Arabian peninsula, and several hundred such species live in Africa, most in small colonies that are widely dispersed due to the difficulty of finding the right food supply and the right habitat.

Female Harvester butterflies (*Feniseca tarquinius*) lay their eggs among woolly aphids (*Neoprociphilus, Pemphigus, Prociphilus,* and *Schizoneura*), and when the larvae hatch, they eat the young aphids, making Harvesters the only carnivorous butterfly larvae in North America. The aphids secrete a sweet substance known as *honeydew* that attracts ants. The ants carefully tend and protect the aphids in order to maintain their supply of nutrients. Because the Harvester larvae are in competition with the ants for available nutrients, they sometimes hide themselves under a mat they spin from silk and then cover with aphid carcasses. Recent research by Youngsteadt and DeVries has found that Harvester larvae can produce a chemical camouflage that mimics the species of aphid on which they are feeding, providing further protection from the normally aggressive ants.

When the Harvester caterpillar is mature and ready to form a pupa, it covers itself with dead aphid carcasses and gets aphid secretions all over the long hairs on its body. The adult butterfly eats only the watery waste excretions of the live aphids, and rarely even alights on a flower.

One type of moth spends virtually its entire life underwater. Used as a biological control of invasive plants, as reported by Coombs, the larvae of the Water Veneer moth (*Acentria ephemerella*) feed on various submersed aquatic plants such as pondweeds (*Potomogeton* spp.) and Canadian waterweed (*Elodea canadensis*) in canals, ponds, and lakes throughout the temperate zones of Europe and much of North America. The wingless form of the female mates on the surface with the fully winged males. The winged female form comes to the surface and floats when she is ready to attract a mate, and she emerges from the water only for a brief courtship flight. After mating, the female dives into the water and lays a clutch of 100 to 300 eggs along the leaves of submersed plants. Larvae hatch from the eggs and burrow inside parts of the leaves, where they feed and build light, water-filled temporary shelters. The larvae may overwinter underwater or inside plant stems, and when they are ready to pupate, they spin heavy, air-filled cocoons.

About 200 butterfly and moth species are predatory. Extraordinary flesh-eating moth caterpillars, *Hyposmocoma molluscivora*, have recently been discovered by Rubinoff in Hawaiian rainforests. They are a small species of case-bearing moth that carries or drags around a silk case that eventually becomes a cocoon. Most of the caterpillar's body is hidden inside the twig-like case as it crawls around. The caterpillar approaches a snail, pokes it to make sure it is alive, and then proceeds to tie it down with silk webbing. After the caterpillar has dined on the immobilized snail, it sometimes takes along the shell, perhaps as camouflage.

Question 5: Does a caterpillar have a skeleton?

Answer: A caterpillar does not have bones or an internal skeleton. It has a fairly hard covering on the outside of its body called an *exoskeleton*, and a butterfly's body is built the same way (see Chapter 2, Question 1: Does a butterfly have bones?). In order to grow, the caterpillar must shed this exoskeleton and grow a larger one, usually doing this four or five times until it pupates (see Chapter 6, Question 6: How does a caterpillar grow?).

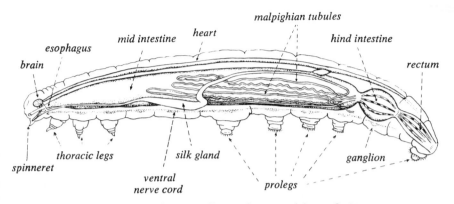

Figure 12.   Major internal organs of larva of a Monarch butterfly (*Danaus plexippus*). Note the lack of an internal skeleton. *(Drawing courtesy of William H. Howe, from* Butterflies of North America, *page 6)*

Question 6: How does a caterpillar grow?

Answer: The caterpillar stage is the time when major growth occurs, although there is growth in the egg stage as well. Growth is regulated by bursts of hormones, and caterpillars grow rapidly and increase in size enormously between hatching from the egg and pupating. If a human baby grew at the same rate, it would be as big as a bus in a month.

After hatching from its egg, the caterpillar starts eating immediately if there is food available, often making its egg case its first meal. Caterpillars excrete copious amounts of waste (*frass*) because of their fibrous diet, and they are often surrounded by large pellets of frass when you find them outdoors. Some species of caterpillars that hatch as winter is approaching enter diapause (hibernation) until spring arrives with a new crop of food plants. The Red Admiral larva (*Vanessa atalanta*), for example, curls the base of a leaf into a tube, secures it with silk, and ties the base of the leaf firmly to a twig. Then it crawls into the tube and remains in diapause until the spring, when the larva emerges and resumes its development. All adult butterflies emerge (*eclose*) from their pupae fully grown.

Like all insects, butterflies have exoskeletons (see Chapter 2, Question 1: Does a butterfly have bones?), so in order to grow, the caterpillar must shed its skin (actually its exoskeleton) several times. The old exoskeleton separates from the inner skin, and fluid forms within this space and partially dissolves, thins, and softens the old exoskeleton from the inside. Meanwhile, a new soft skin forms underneath, called a *cuticle.* The caterpillar swallows air and puffs itself up to split the weakened old skin along a seam and at the same time to stretch the new cuticle so its body has room to grow. The caterpillar remains puffed up until the new skin hardens. Then it releases the air, settles into its new, loose skin, and resumes eating. The caterpillar often eats the discarded old skin.

The caterpillar continues to eat and grow until it needs to molt again. Caterpillars molt an average of four to five times, some up to nine times. Each of these size stages is called an *instar.* The last instar is the pupa (see Chapter 6, Question 7: What is the difference between a chrysalis and a pupa?).

Question 7: What is the difference between a chrysalis and a pupa?

Answer: The words are often used interchangeably to describe the stage in the life of a butterfly or moth during which metamorphosis occurs. When the butterfly caterpillar is ready to molt for the last time, it fastens itself to a branch or other quiet place with a strand of silk. A liquid is secreted from a gland near its mouth, and it becomes a silk thread when it is exposed to the air. Then the final caterpillar exoskeleton is shed and the *chrysalis* is revealed underneath, totally enclosing the head and body. Instead of becoming a new, larger caterpillar, the caterpillar begins to change within the chrysalis and the pupal stage begins. A moth caterpillar first spins a silken cocoon around itself, and then it pupates inside the cocoon (see Chapter 6, Question 10: What is a cocoon?).

The chrysalis is made up of several layers, of which the cuticle is the outer layer or new exoskeleton. The cuticle has an inner layer that contains up to 250 pairs of alternating clear and dense strata. Its shape and density camouflage and protect the insect inside.

Question 8: What happens inside the pupa?

Answer: The pupa is totally enclosed, so the animal within cannot eat, but there are tiny holes in its sides through which it obtains oxygen (see Chapter 2, Question 2: How does a butterfly breathe?). The stored body fat from the caterpillar stage provides the energy needed to accomplish metamorphosis, and a delicate balance of hormones and genetics controls the entire process.

During pupal life, most of the caterpillar's cell membranes break down and their contents are recycled. If the caterpillar had become toxic by eating the leaves of toxic host plants, the toxic chemicals are stored in the hemolymph. During metamorphosis, so that the animal does not poison itself (*auto-toxicity*) when the cell membranes break down, the poison usually goes into the waste fluid (*meconium*), which is released when the butterfly emerges. In some species the toxins are retained but are sequestered during this stage in such a way that they are not auto-toxic.

Very early in the caterpillar's life, several clumps of cells were set aside in different parts of its body. These clumps are called *imaginal discs,* appropriately named since another word for adult is *imago.* Each pair of discs is composed of the cells that are the precursors or forerunners of adult structures: two eye discs, two wing discs, six leg discs, etc. These discs have no function in the caterpillar, but become very active during the pupal stage.

The recycled cell material is used to build a rapidly increasing number of cells in each imaginal disc. Then each disc unfolds and turns inside-out, revealing the adult shape of a wing or a leg or an eye. These emerging regions get organized into the adult shape within the pupa. Some internal body parts, such as the nervous system, are retained and reconfigured; other body parts that are unique to the caterpillar are discarded (see Chapter 6,

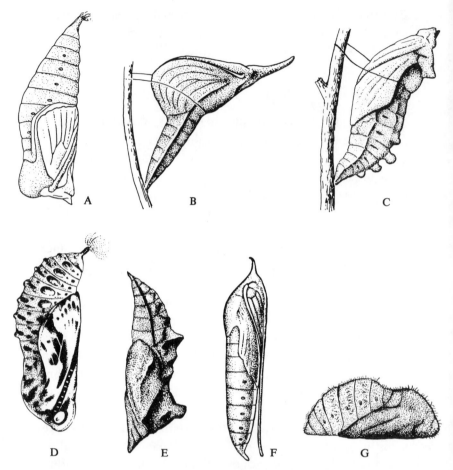

Figure 13. Pupae of various butterflies. A, *Euptychia mitchellii* (after Hubbell);
B, *Phoebis sennae* (after Christman); C, *Battus philenor* (after Christman);
D, *Euphydryas chalcedona* (after Essig); E, *Polygonia satyrus* (after Scudder);
F, *Calpodes ethlius* (after Chittenden); *Callophrys henrici* (after Downey).
(Drawing courtesy of William H. Howe, from Butterflies of North America,
page 9)

Question 9: What happens when the butterfly is ready to emerge from its chrysalis?). Once all the necessary changes have taken place and environmental conditions are favorable, the butterfly is ready to emerge from the chrysalis and stretch its wings.

Question 9: What happens when the butterfly is ready to emerge from its chrysalis?

Answer: The emergence (*eclosion*) of the adult butterfly from the chrysalis is triggered by genetics and hormones in combination with environmental cues. Warmth is an important cue for species that have spent the winter as pupae. Increased humidity is a cue in a desert environment, where rainfall indicates the likely availability of food plants. In a stable environment, eclosion may simply occur when metamorphosis is complete, often at relatively the same time of day for members of the same species.

When eclosion is triggered, a hormonal burst occurs, the butterfly starts to squirm, and the pupal cuticle splits open. Muscular activities begin that allow the adult to quickly emerge. When the adult emerges, a green, reddish, or brownish liquid (*meconium*) spills out. This liquid contains wastes and cells from leftover parts of the caterpillar, such as the five pairs of discarded prolegs (see Chapter 2, Question 8: How many legs does a butterfly have?).

The butterfly is quite vulnerable to predators when it has just eclosed because it is unable to fly. It sits quietly for a while, pumping body fluid (*hemolymph*) into temporarily functional veins in its wings so that the wings can expand and it can fly away to seek food and to find a mate.

Question 10: What is a cocoon?

Answer: A cocoon is a hard, protective outer wrapping made by a moth caterpillar using silk produced from glands near its mouth. Butterfly larvae do not make cocoons. Sometimes the caterpillar will fasten leaves together with silk to form the cocoon,

Figure 14. A Monarch butterfly (*Danaus plexippus*) emerging from its pupa. Even before the insect emerges, the wings can be seen through the pupal covering. (*Drawing courtesy of William H. Howe, from* Our Butterflies and Moths, *page 75*)

while other species form the entire cocoon from silk. When the cocoon is finished, it provides warmth and camouflage. Inside the cocoon the last exoskeleton of the caterpillar molts, revealing the chrysalis, and the pupal stage begins. Sometimes if you look inside an empty cocoon you can see both the remains of the last caterpillar exoskeleton and the empty chrysalis.

When some moth caterpillars have finished growing, they burrow underground and pupate safely there. Some Case moth caterpillars (Psychidae) build a case around themselves that they always carry with them and eventually use as a cocoon. It is made of silk and pieces of plants or soil.

Question 11: How does a moth get out of its cocoon?

Answer: A cocoon is usually quite hard, and when the adult moth is ready to emerge, it uses one of several techniques. Some caterpillars—for example, one of the members of the group known as Jumping Bean moths (*Laspeyresia saltitans*)—prepare in advance by chewing a weakened circle in the cocoon before they pupate so that all the adult moth has to do is push the circle out when it is ready to emerge. Other moths have a file-like organ with which to cut their way out. Most moths produce a liquid when they are ready to break free which softens the cocoon's walls so that they can push their way out.

Question 12: What is a silkworm?

Answer: All caterpillars produce silk from their salivary glands—they lay down a track of silk as they walk and cling to it. They anchor themselves with the silk and may use it to join leaves together to make a shelter.

The finest quality silk is produced by the Silkworm moth (*Bombyx mori*) from Asia, which may spin a single thread about half a mile (0.8 kilometers) or more in length to make its cocoon.

For many years the export of silkworms or their eggs from China was punishable by death in order to protect the secret of silk production. Silkworms have been cultivated for 5,000 years and can no longer survive in the wild because their wings have evolved so that they are too small to allow the moth to fly. The caterpillars cannot move very well and have to be placed directly on their food plant, preferably with the leaves finely shredded.

More than 25,000 cocoons are needed to make one pound of silk. So that the pupae do not damage the silk by trying to emerge, most are killed by boiling the cocoons. The insect is usually eaten after the cocoon has been unraveled. The adults that are kept for breeding have virtually no mouthparts; they do not eat, and die soon after mating and laying eggs.

Question 13: What does a jumping bean have to do with moths and butterflies?

Answer: A so-called jumping bean is actually not a bean at all. It is a small, thin-shelled section of a seed capsule that contains the larva of a small, gray moth called a Jumping Bean moth (*Cydia saltitans*), also called *Carpocapsa saltitans* or *Laspeyresia saltitans*—*carpo* indicating that it lives within a seed, and *saltitans* referring to its jumping behavior.

The larvae feed on a shrub (*Sebastiana pavoniana*) native to the rocky desert areas of Mexico and to a few areas of California. Females lay their eggs on the immature ovaries (capsules) of the shrubs' flowers. After a few weeks the eggs hatch and the tiny larvae eat their way into the flowers' capsules. As the shrubs mature, the capsules become hard and change to a brownish color.

The moth larvae, now trapped inside their new homes, begin eating the developing seed inside the capsules. In the spring, the capsules fall to the ground, and the capsules that contain larvae begin to "jump." When you place a capsule near heat, the larva twists and jumps, thereby causing the jumping movements of the capsule. Perhaps the purpose of this activity is to move the capsule out of direct sunlight, or perhaps there are other explanations for the behavior. Eventually the larva spins a cocoon inside the capsule and the process of metamorphosis takes place.

Question 14: What is the lifespan of a butterfly including all its stages?

Answer: Each species has a relatively set lifespan, spent for a predictable amount of time in each stage of the life cycle, but the typical length of these stages for each species varies widely.

In stable climates, there is a constant supply of food plants for the larvae when they hatch from their eggs, so a shorter life cycle is possible because the adult can lay her eggs soon after fertilization with confidence that her larvae will have enough food. The larvae hatch from the eggs in a few days, grow and pupate efficiently, and healthy adults soon emerge and are ready to mate. Tropical butterflies such as Snouts (*Libytheana carinenta*) and Lyside Sulphurs (*Kricogonia lysides*) mature in around two weeks. Yucca moths (*Tegeticula* spp.) live for no more than two days as adults, so males and females must emerge virtually on the same day so that they can successfully reproduce.

In unstable climates or inhospitable conditions, the female may not lay her eggs at all if she does not find healthy food plants. The larvae may delay hatching from the egg, or the butterfly may stay in the pupa for months or even years if the climate is not suitable. This waiting period is called *diapause,* and it may occur during periods of extreme cold (*hibernation*) or extreme heat or drought (*estivation*). Jogar studied pupae during diapause and observed changes in heart rate, increased water loss, and slowed respiration. Changes in air temperature and humidity can stimulate a break in diapause. Changes in the length of the day typically stimulate emergence from the pupa (*eclosion*). The Arctic dwelling Polaris Fritillary (*Boloria polaris*) requires two years to complete its life cycle in the northern most parts of its range. The extreme cold results in a long period of diapause, so these butterflies are only seen every other year.

Diet plays a role in how long development takes. Larvae that eat tough grasses and other hard-to-digest foods grow more slowly than species that eat easily digestible foods. Bark-eating moths, for example, are tiny and slow-growing. They can take up to four years to complete their life cycle (see Chapter 6, Question 4: What do caterpillars eat?).

A dramatic example of variability occurs with Monarch butterflies (*Danaus plexippus*). Several generations of Monarchs are born each year. Each generation only lives for a few weeks, with the exception of the Monarchs that are born in the fall. This group is biologically and behaviorally different from the other generations, having been instinctively prepared for the strenuous migration to a warmer climate, which can involve a flight of over 2,500 miles (4,022 kilometers). These migratory monarchs can live for six months or more. (See Chapter 8, Question 7: What allows the migrating generation of Monarchs to live so long?)

If a species has one generation a year, or a one-year life cycle, they are said to be *univoltine.* If they have two generations a year, they are *bivoltine,* and if they have many short-lived generations in a year, they are *multivoltine.* When their habitat is harsh, it may take two years to complete one generation, and that is referred to as *hemivoltine.*

SEVEN

Dangers and Defenses

Question 1: Do people eat butterflies and moths?

Answer: Yes, they certainly do. Eating insects, known as *entomophagy*, has been practiced all over the world for centuries because many insects are rich in protein and good sources of vitamins, minerals, and fats. They are tasty, inexpensive, and clean in their own eating habits, most feeding only on fresh, green plants.

The Witchetty Grub is the caterpillar of the large cossid moth (*Endoxyla leucomochla*). It feeds on acacia stems and roots and lives inside them. The grubs are harvested by digging and chopping up the roots. Eaten both raw and cooked by Aboriginal Australians, they are compared in taste to eggs. It is said that ten large grubs provide all the calories, protein, and fat that an adult human needs in a day. Wounds and burns are also treated with a layer of crushed Witchetty Grub, whose oils probably function like a soothing ointment. The Cocopa Indians in Arizona are reported to eat the large White-lined Sphinx moth (*Hyles lineata*). The entrails of the moth are squeezed out, pit-baked, reboiled, parched, and then eaten.

Mopane moth caterpillars (*Imbrasia belina*) are harvested in large quantities by some ethnic groups in southern Africa, especially in Botswana and Zimbabwe. The caterpillars are gutted and then baked. Silkworm moth pupae (*Bombyx mori*) are considered delicacies in China, and are eaten fried in Korea and Japan. The larvae of the Yucca Giant-Skipper (*Megathymus yuccae*) are edible and used in the Mexican dish "Gusanos de Maguey." The people of Zaire eat more than 35 different types of caterpillars, usually toasted or sautéed in butter.

Aboriginal Australians harvested Bogong moths (*Agrotis infusa*) when the moths gathered to rest (*estivate*) during the hot, dry season. The moths were cooked in hot ashes to burn off their wings and legs. Their heads were removed, and the remaining bodies could be eaten as is, or ground into a paste and kneaded into cakes and then roasted. The moths were an important source of nutrients during the hot season, when there was very little to eat. It is said that those who ate Bogong moths for the first time usually had bad stomach cramps and vomiting.

Question 2: What dangers do caterpillars face?

Answer: It is not unusual for only one or two caterpillars out of a hundred to survive to adulthood. Caterpillars are voracious eaters and cause a lot of destruction of crops and vegetation, but they have a lot of natural enemies that kill them and thus reduce the amount of damage they can cause. Wagner thoroughly discusses the natural enemies that attack them, including predators, parasites, parasitoids, and pathogens. Some of the animals that prey on caterpillars are birds, spiders, ants, mice, lizards, dragonflies, and even large mammals like bears and foxes.

Caterpillars are very sensitive to the environment. Wet conditions stimulate the growth of fungi that kill caterpillars, and very dry conditions can lead to dehydration (*desiccation*), which can be fatal. Marsh Fritillary larvae (*Euphydryas aurinia*) have an interesting way of overcoming the limitations of their cool, damp habitat. They spin a communal web near large patches of their food plant, and being close together is thought to raise their body temperatures (see Chapter 2, Question 14: Why are butterflies called "cold-blooded"?).

The consequences of natural habitats changing to a human-dominated landscape can be lethal to caterpillars, butterflies, and moths. Deforestation, large, brightly lit developments, airports, pesticides, etc. are some of the changes that endanger Lepidoptera (see Chapter 9, Question 3: Why don't I see as many butterflies as I used to?).

Many plants have defenses that minimize their appeal to hungry larvae. Two plants of the Asteraceae family are phototoxic, which means light induces the production of toxic chemicals. The Redbanded Leafroller moth (*Argyrotaenia velutinana*), for which these are host plants, avoids direct exposure to the sun while on the plant. The females lay their eggs in the shady parts of the plant, and the larvae hide and spin opaque shelters. Larvae of the Blackberry Looper moth (*Chlorochlamys chloroleucaria*) that also feed on these plants prefer to eat pollen, which is a nonphototoxic tissue of the plant.

Some plants have toxic chemical barriers in the leaves, so that only certain larva can eat their leaves and survive, often becoming toxic themselves (see Chapter 7, Question 3: How do caterpillars defend themselves?). Other plants give off bouquets of volatile chemicals that repel insects or attract predators of the species that use the plant as a host.

Corn plants, one of the favorite foods of the Cottonworm moth larva (*Spodoptera littoralis*), release a chemical distress signal when they eaten by the caterpillars. The chemicals attract adult female parasitic wasps (*Cotesia marginiventris*) that attack by laying a single egg in the posterior of the larva. When the wasp egg hatches, it burrows into the larva and eats its insides. The well-fed wasp grub then burrows out of the larva and spins a cocoon attached to the larva's back in which to pupate.

Parasites coexist with their host, feeding on it but not killing it immediately. Some parasitic wasps lay their eggs on a plant and their larvae wait to ambush butterfly larvae that come to the plant to feed. They attach themselves to the skin of the butterfly larvae and bore inside the insect. Other parasitic insects use a needle-like ovipositor, an egg-laying appendage, to break the skin and lay their eggs inside the host's body. Some wasps lay eggs into pupae, and a very tiny species lays eggs into the eggs of butterflies and moths.

One parasitic wasp species, *Ichneumon eumerus*, found only in southern France and northern Spain, uses a rather elaborate tactic to victimize a species of butterfly larvae that live among ants. After feeding within the buds of gentian flowers and growing through several instars (see Chapter 6, Question 6: How does a caterpillar grow?), larvae of the Alcon Blue butterfly (*Maculinea rebeli*) drop to the ground. The butterfly larvae emit an odor identical to that of the larvae of the red ant (*Myrmica schencki*), so they are picked up by the adult ants, taken back to the nest, and placed among the ant brood. Thomas reports that the female *Ichneumon eumerus* wasp locates an ant colony that contains a butterfly larva, then enters the nest and introduces pheromones that cause the ants to become agitated and aggressive to such a degree that they attack each other. In the ensuing chaos, the wasp slips in unnoticed and lays an egg on the butterfly larva. The ants continue to feed the butterfly larva, and the wasp grub emerges from its egg and burrows into the body of the butterfly larva, feeding on it and growing inside it. When the butterfly larva eventually pupates, the developing wasp eats the rest of its host, and a wasp, not a butterfly, emerges from the pupa.

Parasitoids are smaller than their prey, and they differ from true parasites in that they nearly always kill their host, usually very slowly. They feed on nonessential tissues and organs as long as possible, allowing the caterpillar to behave normally and avoid larger predators that might endanger host and parasitoid. When the food supply is depleted, they continue to eat until they kill the host.

Pathogens are microscopic bacteria, fungi, and viruses. They are the most common killers, attacking larvae and weakening or killing them. There are some bacteria (*Wolbachia* spp.) that kill only males.

Question 3: How do caterpillars defend themselves?

Answer: Caterpillars have many defenses, often more than one. When endangered, some caterpillars bite, others regurgitate sticky, smelly fluids, others make noises, drop to the ground, or wiggle violently.

The Walnut Sphinx moth (*Amorpha juglandis*) makes a hissing sound by forcing air out of its spiracles. The Abbott's Sphinx moth (*Sphecodina abbottii*) squeaks like a mouse. When some Swallowtail larvae are disturbed, they rear up on their back legs and expose a pair of fleshy hornlike structures called *osmeteria*, which are part of a y-shaped organ that is otherwise hidden in a pocket behind the head. The osmeteria extend and swell like pushing out the fingers of a glove. They can also emit a foul odor, all of which discourages predators or at least misdirects their attack. The White-lined Sphinx moth larva (*Hyles lineata*) assumes a threatening posture and emits a thick, green substance from its mouth.

Caterpillars often have hairs, spines, or barbed hooks. Generally these are for show and are quite harmless, but there are a few stinging caterpillars, like those of the arctiid moth (*Pareuchaetes pseudoinsulata*), which have hairs that can cause a burning sensation when touched. These hollow, quill-like hairs are connected to poison sacs. When the hairs are touched, they break through the victim's skin, releasing the poison. Reactions to these and other highly toxic caterpillars, such as the Puss moth caterpillar (*Megalopyge opercularis*), vary according to the sensitivity of the victim and the location of the sting. Foot reports that mild itching, burning, swelling, nausea and vomiting, and fever in children are typical.

In addition to manufacturing their own defensive substances to achieve protection from predators, many species of larvae accumulate (*sequester*) unpalatable or toxic substances from their host plants. Longwing caterpillars (*Heliconius* spp.) feed on the passion flower vine (*Passiflora auriculata*), which contains potentially toxic compounds, and these caterpillars can become toxic to predators. Enzymes in the caterpillar's digestive system convert the compounds to toxic cyanide when a predator's bite releases the enzymes. Toxic larvae usually have bright, warning (*aposematic*) coloration.

Question 4: What dangers do butterflies face?

Answer: Thousands or possibly millions of butterflies are killed by cars as they swarm along highways in warm weather. Butterflies and moths in all their stages are food for birds, lizards, spiders, wasps, beetles, ants, shrews, toads, mice, and even some larger animals. North Indian Rhesus monkeys eat butterflies and moths like cookies, and the Mastiff bat (*Molossus pretiosus*) dines exclusively on moths. Viruses and other diseases, extreme temperatures, wind, drought, fire, floods, pesticides, pollution, and deforestation are additional dangers (see Chapter 9, Question 3: Why don't I see as many butterflies as I used to?).

Question 5: How do butterflies defend themselves?

Answer: Different species have different ways of defending themselves. Some butterflies make startling noises (see Chapter 3, Question 10: Do butterflies ever make noises?). Some butterflies have scales that emit smells to repel predators. Some brightly colored butterflies have drab underwings, and they frighten or confuse predators by flashing their colors and patterns. And the slippery, dust-like scales on a butterfly or moth make it easier for them to slip out of a spider's web than it is for an insect that has only a hard exoskeleton.

The shape of the wing helps protect some species. Many swallowtail varieties have a long tail at the rear end of each wing. If the tempting tail end of the wing is attacked by a predator, it easily tears off and the butterfly escapes.

Some nontoxic butterflies and moths use fast, erratic flight to evade predators. Some moths stop flying and go into a rapid free fall, dropping quickly to the ground when they feel threatened. Some animals mimic the movements of other species that are ignored by visual predators such as lizards and birds because they are too difficult to capture (*evasive prey mimicry*). Metalmark moths (*Brenthia* spp.) mimic jumping spiders, one of their own predators.

Other butterflies have "cryptic" or camouflage patterns, which help them blend in with the bark or leaves on which they rest. The Indian Leafwing butterfly (*Kallima inachus*) looks so much like a leaf when it is at rest that only very close inspection reveals that it is actually a butterfly.

When there are predatory bats in their habitat, nocturnal Noctuid moths exhibit an *acoustic startle response* when they sense a high-frequency echolocation call from a bat. Their response is to move evasively or to drop to the ground. Some moths have another defense against bats. They have tufts of hairs (*setae*) that give the moth a fuzzy appearance and prevent the moth from reflecting bat sonar, thereby preventing the bat from locating the moth.

Figure 15. Common Mormon Swallowtail (*Papilio polytes*) missing one tail, suggesting a predator was drawn away from the vulnerable body. (*Photograph by Carol Butler*)

Figure 16. An Indian Leafwing butterfly (*Kallima inachus*) simulates a dead leaf. (*Drawing courtesy of William H. Howe, from* Our Butterflies and Moths, *page 167*)

Question 6: Are butterflies poisonous?

Answer: Bright colors or patterns (*aposomatic* coloration) usually indicate that a butterfly is toxic. Monarch butterflies ingest vomit-inducing cardiac glycosides and bitter tasting, toxic cardenolides from milkweed plants during the larval stage, and the toxins are sequestered in areas of the butterfly's body in different

How Do Butterflies Defend Themselves?

Camouflage that blends in with their surroundings.
Bright colors that flash and startle predators.
Flying fast and/or flying in a zigzag pattern.
Becoming toxic to predators by eating toxic food plants as a caterpillar.
Resembling a toxic butterfly to confuse predators.
Making noises to startle predators.
Giving off bad smells.

concentrations. Depending on where a predator bites the butterfly, the predator will vomit or will certainly find the butterfly distasteful. There are, however, some predators who are not bothered by the toxins (see Chapter 7, Question 7: Do toxic butterflies have any predators?).

Male Paper Kites (*Idea leuconoe*), found in the wet evergreen forests of East Asia, require nutrients from certain plants that they use to synthesize pheromones. A chemical in one of these substances, danaidone, is poisonous, and it is passed to the female in the sperm packet when the butterflies mate. This substance ends up in the tissues of the female and in her eggs, protecting both from predation.

Batesian mimics are nontoxic species that closely resemble toxic butterflies living nearby. Predators learn to avoid the toxic *models* after tasting one individual and violently vomiting, and the nontoxic mimics are generally avoided as well. Batesian mimics usually emerge from their chrysalides later in the season than their toxic neighbors, after the birds have already learned to avoid their "type."

Mullerian mimics are species that *are* actually toxic and have evolved to resemble other toxic species. The similarity in appearance means that predators learn more quickly to avoid them. The Viceroy butterfly (*Limenitis archippus*) mimics its toxic neighbors: it resembles the toxic Monarch (*Danaus plexippus*) in habitats where Monarchs are plentiful, but in the American south and southwest, where Monarchs are relatively rare, the Viceroy is typically darker and closer in color to the toxic Queen butterfly (*Danaus gilippus*).

Mimicry has a mutual benefit for the model and the mimic, because both species share the number of sacrificed butterflies necessary to teach predators that all similar butterflies taste bad and should be avoided.

Question 7: Do toxic butterflies have any predators?

Answer: The answer is yes, and because there is more research about Monarchs than about any other butterfly, we can give specific examples of toxin-resistant Monarch predators, and of course it is probable that there are toxin-resistant predators to other species as well. The parasitic protozoan *Ophryocystis elektroscirrha* is transmitted from generation to generation by spores that are shed from the scales of the adult female onto her eggs and onto the surface of the milkweed plants on which she lays eggs. The spores are also transferred between females and males during mating.

The scansorial (climbing) Black-eared mouse (*Peromyscus melanotis*), common around the Monarch's overwintering sites in Mexico, eats the body of the Monarch but not the wings. Glendinning reported that, on average, mice ate about 40 freshly-killed Monarchs in a typical night.

Two birds, the Black-headed Grosbeak (*Pheucticus melanocephalus*) and the Black-backed Oriole (*Icterus abeillei*), have been found by Brower to be responsible for killing an estimated 44 percent of the Monarch population at their Mexican roosting sites. Although the Orioles vomit after consuming small amounts of the toxins sequestered by Monarchs, they are usually able to avoid poisoning by not eating the butterfly's cuticle (its skin), which is where Monarchs store the toxins. Instead, they slit open the body and strip out the soft insides.

The Grosbeak is relatively insensitive to the toxins. It eats the entire butterfly, though it seems more likely to prey on males, which tend to be less toxic.

Question 8: Why do some butterflies have eye-like spots on their wings?

Answer: The large Owl butterfly (*Caligo* spp.) earned its name because it has large spots that resemble eyes toward the edges of the ventral surface of its wings (the side you see when the wings are closed.). Butterflies with eyespots are frequently found with bite marks on the edges of their wings. The eyespots seem to draw the attention of predators, either because their instinct is to attack the eyes of their prey, or simply because the markings are conspicuous. Predators are misdirected away from the butterfly's vulnerable body, and even with pieces of the wing missing, most butterflies are still able to fly.

Researchers in Scotland recently did an interesting experi-

Figure 17.   An Owl butterfly (*Caligo atreus*) showing an eyespot and wing damage near the eyespot area. (*Photograph by Carol Butler*)

ment with the Peacock butterfly (*Inachis io*), another butterfly with prominent eyespots. They captured butterflies and masked their spots and then re-released them in the wild. When the eyespots had been masked, most of the butterflies were caught by birds in short order. All but one of the control group of butterflies with unmasked spots survived.

Question 9: Can a butterfly or moth harm me?

Answer: Although some butterflies are toxic to predators, they cannot bite you or expose you to their toxins in the course of normal handling. However, if *you* bite a toxic butterfly, you might, at the very least, find it quite distasteful.

Handling toxic caterpillars is dangerous, however. Unlike butterflies that only have curled proboscises and lack jaws, caterpillars have mouth parts and can bite if threatened. They have other defenses as well (see Chapter 7, Question 3: How do caterpillars defend themselves?).

Mäkinen-Kiljunen confirmed a baker's report of a respiratory allergy to the Mill moth, also known as the Mediterranean Flour moth (*Ephestia kuehniella*), and Popa reported allergies to the Angoumois Grain moth (*Sitotroga cerealella*). These conditions seemed to have developed over years of exposure to the moths.

In 1980, the U.S. Department of Agriculture asked the National Institute for Occupational Safety and Health (NIOSH) to conduct a health hazard study to assess the prevalence of symptoms associated with raising colonies of insects in confined spaces. Thirty-three percent of the entomologists and laboratory technicians working directly with insects reported what seemed to be allergic symptoms, compared with 9 percent of workers who had little or no direct contact with insects. Two-thirds of the implicated insects were in the Lepidoptera order. The most common symptoms included sneezing or runny nose (73 percent), eye irritation (68 percent), skin irritation or skin rash (41 percent), cough (38 percent), wheezing (26 percent), and shortness of breath (24 percent).

Evo Devo: Seeing Spots Before Your Eyes

The development of various body parts in unrelated animals was historically thought to have evolved in entirely different ways, but their development has been found to be governed by the same genes. Scales on a butterfly's wing have evolved as modifications of the sensory bristles on the wings of insects like the fruit fly, and the development of both the scales and bristles are controlled by the same gene. More amazingly, the gene that controls the formation of an arm on a human is the same gene that controls the formation of a wing on a bird, a fin on a fish, and the eyespots on a butterfly's wing.

An early example of this work was reported in 1980 by Fred Nijhout of Duke University. He studied the normal locations of eyespot patterns on butterfly wings. In his lab, working on the pupal stage, he was able to kill a tiny patch of cells in the spot where the center of an eyespot would normally be expected to appear, and when the butterfly *eclosed* (emerged from the pupa), no eyespot had formed in the expected location. When the small group of cells was transplanted to a different place on the pupal precursor of the wing, a new eyespot appeared at the predicted location on the wing when the butterfly eclosed.

Each wing develops from a flat disc of cells in the caterpillar. Subsequent research by others found that about a week or so before the butterfly ecloses, certain proteins that mark the location of the eyespot pattern can be observed, although the wing is still a tiny immature disc. These so-called tool kit proteins produce substances called *morphogens* that have the capacity to influence the development of cells. The morphogens have been demonstrated to be active at particular locations, and they "flip a switch" in the DNA that tells the genes what to do, for example, to make an eyespot.

Carroll discusses the research that is being done to understand the central principles governing the emergence of forms and patterns in the developing embryo. This new area

(continued)

Metamorphosis, three stages in the life cycle of the Blue Morpho (*Morpho peleides*)

The Blue Morpho caterpillar has tufts of irritating hairs for protection from predators. *(Photograph by Hazel Davies)*

Blue Morpho pupae ready for shipping from a Costa Rican butterfly farm. *(Photograph by Hazel Davies)*

Scales covering the wings reflect light, giving the Blue Morpho an iridescent appearance. *(Photograph by Hazel Davies)*

Caterpillars

The larvae of the White-lined Sphinx moth (*Hyles lineata*) have a varied diet, feeding on plants in the evening primrose and apple families in addition to various herbs and woody species. (*Photograph by Hazel Davies*)

Found only in southeastern Florida, the brightly colored larvae of the Atala Hairstreak (*Eumaeus atala*) have a pungent odor. (*Photograph by Hazel Davies*)

This Tersa Sphinx moth larva (*Xylophanes tersa*) is parasitized with Braconid wasps. A wasp will emerge from each tiny white cocoon, having grown as larvae by feeding on the caterpillar. (*Photograph by Hazel Davies*)

Larvae of the Zebra Longwing (*Heliconius charitonius*) feed only on the passion flower vine (*Passiflora spp.*), from which they sequester toxins. (*Photograph by Hazel Davies*)

Drinking

A Longwing butterfly (*Heliconius* sp.) with rolled proboscis. *(Photograph by Carol Butler)*

A Great Southern White (*Ascia monuste*) uncurls its proboscis to sip nectar from Pentas blooms. *(Photograph by Hazel Davies)*

A Hecale Longwing (*Heliconius hecale*) collects pollen on its proboscis. *(Photograph by Hazel Davies)*

An "Eighty" butterfly (*Diaethria candrena*) drinking mist from a waterfall in Argentina. *(Photograph by Hazel Davies)*

Dimorphism

The Green Bird-wing (*Ornithoptera priamus*), like many butterflies, is sexually dimorphic, which means that the males and females look very different. The male has iridescent green markings, while the female is larger but only has a tiny stripe of the green on her thorax. (*Photographs by Hazel Davies*)

The Eastern Tiger Swallowtail (*Papilio glaucus*) has two forms of the female. The yellow form resembles the male of the species, and the dark form has only a shadow of the tiger stripes and is more common in the southern part of the range, where it is thought to mimic the toxic Pipevine Swallowtail (*Battus philenor*). (*Photographs by Hazel Davies*)

The Indian Leafwing (*Kallima inachus*) is a remarkable leaf mimic and is well camouflaged when resting with its wings closed, but flashes startling blue and orange when the insect takes flight. *(Photograph by Carol Butler)*

The beautiful Olivewing (*Nessaea aglaura*) has excellent camouflage with its leafy coloration on the ventral surface of the wings. The dorsal surface is bright, and when flashed at a would-be predator, might startle it for a few seconds, long enough to allow the butterfly to escape. *(Photograph by Carol Butler)*

The brown ventral (underside) of the Blue Morpho butterfly is a sharp contrast to the iridescent dorsal (topside) of the wings. *(Photograph by Carol Butler)*

Skippers and Snouts

The Ocola Skipper (*Panoquina ocola*) was photographed in south Texas on honey mesquite. (*Photograph by Carol Butler*)

The Long-tailed Skipper (*Urbanus proteus*) is common in southern Florida, Texas, and Mexico. (*Photograph by Carol Butler*)

The tiny Eufala Skipper (*Lerodea eufala*) is common in the southern United States and in Mexico. (*Photograph by Carol Butler*)

Snout butterflies (*Libytheana carinenta*) mating. Their elongated mouthparts (labial palpi) and their coloring suggest a resemblance to dead leaves. (*Photograph by Carol Butler*)

Luna moths (*Actias luna*) are large and quite beautiful. They are common throughout the eastern United States, and are seen in southern Canada as well. *(Photograph by Carol Butler)*

Close-up of an Atlas moth face (*Attacus atlas*) showing the feathery antennae characteristic of many moths. *(Photograph by Carol Butler)*

A startled Polyphemus moth (*Antheraea polyphemus*) shows off its eye spots. *(Photograph by Hazel Davies)*

Miscellaneous

The Common Glass-wing (*Greta oto*) has very few scales on its wings, revealing the translucent membrane that underlies the scales on all butterflies and moths. (*Photograph by Carol Butler*)

This is an extremely unusual genetic mix-up called a *gynandromorph*. Its left half is female and its right half is male. *Parnassius autocrator, 1913.* (*Collection of Naturalis, the National Museum of Natural History, Leiden, The Netherlands, photograph by Carol Butler*)

This is a very old specimen still in remarkable condition, identified in 1777 but collected earlier. *Papilio ulysses, 1758.* (*Collection of Naturalis, the National Museum of Natural History, Leiden, The Netherlands, photograph by Carol Butler*)

Evo Devo: Seeing Spots Before Your Eyes (*continued*)

of scientific study is at the interface of embryology and evolutionary biology. Uncovering the mysteries of the process in which a single-celled egg develops into a complex, multi-billion-celled animal is developmental biology, and studying the evolution of the genes involved in this process by comparing how they are expressed in different organisms makes it evo devo.

Ongoing research on sea urchin embryos by Eric H. Davidson at the California Institute of Technology is exploring how a group of genes locks development into a particular path, thus limiting the range of natural selection. A "kernel" of about five genes has been found to be essential in forming the gastrointestinal tract, also called the "gut," referring to the system of organs that control digestion and excretion. This "kernel" is thought to have existed for over 500 million years, across species, and although many genes can be modified or even removed, if any one of this kernel of core genes is changed, no embryo will form.

Question 10: How does an egg defend itself?

Answer: Females usually lay anywhere from 20 to over 1,000 eggs, depending on the species and the environment. But in all species the mothers abandon their eggs once they are laid, and the fathers have abandoned the mothers as soon as mating was concluded.

The eggs face many dangers. Fatouros studied a wasp (*Trichogramma brassicae*) that detects the odor of a mated female butterfly that has been passed an anti-aphrodisiac pheromone by her mate to make her less attractive to other males. The wasp rides on the butterfly until she lays her eggs, and then parasitizes the freshly laid eggs.

Many studies have been conducted and many estimates are offered about what proportion of eggs hatch into healthy larvae, and what proportion of eggs survive to become adult butterflies. Survival estimates range from 1 percent to 29.7 percent in a sampling of studies, but clearly laying a large number of eggs is a good defense because more eggs are likely to survive. Kinoshita's study of the Yellow Tip butterfly (*Anthocharis scolymus*) found that larvae from eggs laid early in the season had a higher survival rate than larvae from eggs laid later in the season because the later larvae were cannibalized by other larvae on the host plant.

Orians discussed how attractive eggs are to predators, and the importance of defensive strategies that help the eggs survive as elements in natural selection. Females often lay their eggs on the underside of leaves or in other hidden places that makes

Figure 18. Parasitic wasps (*Trichogramma brassicae*) on the wing of a female Large White butterfly (*Pieris brassicae*). (*Used with permission of Dr. Nina Fatouros and Dr. Hans M. Smid, Laboratory of Entomology, Wageningen University, The Netherlands*)

the eggs inaccessible to predators. Some females spray their eggs with substances that warn off other females who might be tempted to lay eggs nearby. Other species coat their eggs with poisons or attach toxic hairs to repel predators.

Question 11: How does a pupa defend itself?

Answer: The last instar of a butterfly caterpillar secretes a strand of silk in liquid form. The liquid becomes a strong fiber when it is exposed to the air, and it attaches the pupa to a branch or other quiet place. This defends the pupa against being easily removed from its safe spot. Although pupae are fixed in place, they are very much alive, and when disturbed, some pupae wiggle and jerk violently. In some species, the wiggling produces sounds (see Chapter 3, Question 10: Do butterflies ever make noises?). Bingham reported that the pupae of the Southern Birdwing (*Troides minos*), found in India, makes a loud noise when touched. The noise does not seem to be just from rubbing parts of the pupae case against each other (*stridulation*), but may also involve pushing air through spiracles, to achieve the curious "*pha-pha*" sound. In Costa Rica, the pupae of the Stinky Leafwing (*Historis odius*) are called *pescadillo,* or little fish, because of their tendency to wiggle.

The chrysalis is the skin of the last stage of the caterpillar. The caterpillar is enclosed in the chrysalis during the pupal stage. This skin is fairly dense and tough, and it camouflages and protects the insect inside. Some pupa blend in with the surrounding plants, camouflaged as twigs or leaves. Some have gold bands or gold flecks, and some are silver and beautifully reflective, perhaps a form of warning coloration. Other chrysalides look like bird droppings.

A moth caterpillar first spins a hard cocoon within which it encloses itself, and then it pupates inside the cocoon (see Chapter 6, Question 10: What is a cocoon). The cocoon usually lies on the ground and provides an extra defensive layer for the pupa. Some species form the cocoon underground and remain there for additional protection from predators during the pupal stage.

Question 12: Do butterflies fight?

Answer: Some male butterflies do have competitive contests with other males, but they do not have any traits that are usually associated with animal aggression. They compete for mating opportunities in courtship rituals in which they fly around or dart at each other at high speed.

Skippers (family Hesperiidae) and Hairstreaks (family Lycaenidae) are groups of butterflies that have some members that aggressively defend their territory. The Common Evening Brown (*Melanitis leda*) in tropical Australia was observed for a five month period by Kemp to determine the nature of the males' territorial contests. Nonresident males arrived and perched in an occupied territory until the resident male challenged them and drove them away. These asymmetrical owner-intruder contests were repeated numerous times over several days with the same pairs and the same outcome. Speckled Wood butterflies (*Pararge aegeria*) were studied by the same investigator to try to understand the nuances involved in this type of competition. The individuals who occupied the territory were more likely to triumph in the species studied, regardless of size or age.

Question 13: How do butterflies survive harsh weather?

Answer: Some changes in the environment signal that bad weather is on its way, and the environmental triggers (such as changes in day length or temperature) stimulate many animals and insects to prepare instinctively to survive until conditions are more favorable.

When harsh conditions are coming, some butterflies migrate. They move in a large group to an area with milder weather and more food. Although over 200 species of butterflies in all parts of the world migrate from time to time, sometimes in huge numbers, the Monarchs (*Danaus plexippus*) in North America are the only species with a predictable, annual, regular round-trip migration (see Chapter 8, Question 1: Do all butterflies migrate? and Chapter 8, Question 2: Why do they migrate?).

Some other species enter a state called *diapause* in which they have very little metabolic activity and are quite inactive. It is a state of suspended animation without feeding and in which virtually no growth occurs. This is a deeper state than *hibernation* (lying dormant in the cold) and *estivation* (lying dormant in the heat).

The stage of life in which diapause can occur is genetically determined, and depending on the species it can occur in the embryo, larva, pupa, or adult stage. Among butterflies and moths, pupal diapause is the most common. Some species native to harsh climates remain in diapause for up to nine months. Caterpillars of some arctic and mountain species can take over three years to mature because they remain in diapause for such a long time, and some pupae may be thawed and frozen several times until favorable conditions return. Species that overwinter as adults find sheltered places to hide. Much research has been done on Arctic butterflies. Keven and Shorthouse review some interesting studies on an undated website, http://pubs.aina.ucalgary.ca/arctic/Arctic23-4-268.pdf.

EIGHT

Butterflies on the Move

Question 1: Do all butterflies migrate?

Answer: Not all butterflies migrate, but quite a few species around the world do so, as Larsen observed, often over very long distances and in large numbers. Migration is a predictable, usually seasonally determined, simultaneous movement of individuals in anticipation of climactic changes. In addition to butterflies, many species of birds and whales migrate annually.

Most butterfly migrations probably go unnoticed, except by naturalists and keen observers. But occasionally the number of individuals migrating reaches such epic proportions that no one in their path can miss the giant cloud of insects. They can even be hazardous to traffic, clogging radiators and covering the windshields of cars. Each spring, the mass movement of Purple Crow butterflies (*Euploea* spp.) prompts officials in Taiwan to take action. It is estimated that over 10,000 butterflies per hour fly across a stretch of highway, and many are dragged into the traffic by turbulence. Protective netting to make the butterflies fly higher, ultraviolet lights to encourage them to fly below the elevated roadway, and lane closures are all used to help more of the butterflies reach their breeding grounds.

The migratory flight may be one direction only, moving into better breeding grounds, such as occurs with the Great Southern White butterfly (*Ascia monuste*) in Florida. Or it may be to hibernation or estivation sites, with a return flight by the same individuals after the period of diapause (see Chapter 7, Question 13: How do butterflies survive harsh weather?). This is seen in the Australian Bogong moth (*Agrotis infusa*), which migrates from the plains of New South Wales and South Australia to caves in the Australian Alps to estivate during the summer. In southern India spectacular and large-scale migrations of multiple species of butterflies, including the Dark Blue Tiger (*Tirumala septentrionis*), the Double Banded Crow (*Euploea sylvester*), and the Common Indian Crow (*Euploea core*) are associated with the heavy rains of the monsoon season.

The most well known of all insect migrations—from across the eastern United States and Canada to hibernation sites in Mexico—is completed annually by the Monarch butterfly (*Danaus plexippus*). Our discussions will focus on the North American Monarchs, although Monarchs are found in many other parts of the world, such as in Australia, New Zealand, Hawaii, and other Pacific islands.

Many other species of butterflies travel to populate new areas and to extend their range as reported by Salvato and others. The flights do not have a predetermined direction and are in response to unfavorable changes in local habitat. This movement to find better conditions is called dispersal or emigration. Factors prompting dispersal include population explosion, which can lead to overcrowded conditions, or drought or other unseasonable weather conditions that can result in a shortage of host plants, or human impact, such as deforestation.

Question 2: Why do they migrate?

Answer: Migratory species instinctively migrate in anticipation of predictable environmental changes, such as the onset of winter or a prolonged dry season. They move to another area, where they may go into diapause for a short time until the conditions are suitable for breeding. When Monarchs (*Danaus plexippus*) first arrive at their overwintering sites in Mexico, for example, the weather in their mountainous destination is quite cool (see Chapter 8, Question 5: Do all migrating Monarchs go to the same place?). They enter diapause until temperatures rise and then they begin breeding.

Scientists think the ancestors of Monarchs were tropical butterflies that moved north each spring to take advantage of the abundant milkweed (*Asclepias* spp.) growing during the summer. But having never evolved the ability to tolerate the winters in temperate regions, Monarchs continue to migrate to favorable overwintering sites each year. Scientists do not know how long ago the Monarchs started their annual migration, though it may date back as far as the end of the last ice age 10,000 years ago.

Kishen Das studies migration patterns in southern India. Between March and April, the monsoon season in the Western Ghats mountain range is probably the trigger for multiple species of Danaidae, or Milkweed butterflies, to move to the plains below and then on to the Eastern Ghats. During September and October, when winter arrives, the butterflies move back toward the southwest, to the Western Ghats, to take advantage of the thick green cover. The late winter monsoon in the northeast of Taiwan causes millions of Purple Crow butterflies (*Euploea* spp.) to migrate to warmer, sheltered valleys in the south.

The Australian Bogong moth (*Agrotis infusa*) migrates each spring from the lowlands of southern Queensland, South Australia, and New South Wales to spend the summer at estivation sites in caves in the Australian Alps. Having spent the summer tightly packed together in caves and rocky crevices, the same adults instinctively time their return to the lowlands to coincide with the availability of host plants. The purpose of their migration may be to delay laying eggs when host plants are not available. Although the climate in southern Australia is warm during the summer, and would be favorable to egg and larva development, the lowlands are dominated at that time of year by grasses rather than the wheat, barley, and vegetable crops that are the host plants for the moth caterpillars.

Nonmigratory butterflies have evolved a method to help them survive inhospitable seasonal changes in climate. They enter a state called *diapause* (see Chapter 7, Question 13: How do butterflies survive harsh weather?). Species that enter diapause are genetically programmed to do so at a set time during their normal life cycle, some species as eggs, others as larvae, pupae, or adults.

Question 3: How do scientists study migration patterns?

Answer: The North American Monarch's (*Danaus plexippus*) migration route is so accessible and predictable that scientists have studied it for many years. Several methods have been used to gain insight into this amazing phenomenon, including tagging programs, monitoring programs, the study of isotope patterns, and research into navigation.

The first tagging program was started in the 1930s, at the University of Toronto, by Dr. Fred Urquhart. This study eventually led scientists to the 1975 discovery of the Monarchs' overwintering sites in Michoacan, Mexico. Current tagging programs include Monarch Watch, based at the University of Kansas, and the Monarch Program, based in California. These programs bring together researchers, students, teachers, and thousands of volunteers across the United States and Canada. Participants catch pre-migratory Monarch butterflies and apply a tag, a sticker less than half an inch (1 centimeter) in diameter, to the wing of each Monarch. The tags ask the person finding the tagged butterfly to send it to the research organization. Recovery of tagged Monarchs reveals a great deal about how fast Monarchs fly and the routes they take.

Monitoring programs such as Journey North, Texas Monarch Watch, and the Monarch Monitoring Program collect information on migration without catching and tagging butterflies. Participants report their first Monarch sightings each spring, and this helps build information over several years to compare when and where the butterflies travel and how weather patterns may affect the migration.

Isotopes are different forms of an element, each having a dif-

Figures 19, 20, 21. A small sample of the thousands of tagged Monarch butterflies that made it possible to learn about the annual Monarch migration. (*Figure 19 courtesy of Carol Sherman; figure 20 courtesy of Dale Clark; figure 21 courtesy of Vicky Temple*)

ferent atomic mass. The elements hydrogen and carbon have several different stable isotopes that can be found in nature. Wassenaar studied isotope patterns in Monarchs and found that it is possible to roughly determine from where the butterfly originated. Very basically, different parts of the world have different amounts of the two stable isotopes of hydrogen, and when plants take up water they obtain an isotope pattern that reflects their geographical region. When Monarch larvae eat they inherit the isotope pattern of the milkweed plants (*Asclepias* spp.) on which they fed. A study of the isotope patterns of Monarchs at overwintering sites in Mexico revealed that about half of 597 individuals analyzed had developed as caterpillars in the midwestern United States.

Question 4: How do Monarchs navigate over long distances?

Answer: Scientists have devised numerous experiments to find out more about how Monarchs navigate, such as testing the Monarch's use of ultraviolet light, fooling the butterflies' internal clock into thinking it is a different time of day (see sidebar on circadian rhythms), or working with magnetic fields. Breakthrough research became possible with the development of a butterfly flight simulator in 2001 by biologists Mouritsen and Frost at Queen's University in Ontario, Canada.

The flight simulator apparatus is a large, white, translucent cylinder placed vertically, with a small computer fan blowing a controlled stream of air up through a hole in the bottom. One at a time, each Monarch is tethered to a small mounting stalk made of tungsten wire. Beeswax is used to glue the tether to the rear end of the butterfly's body, and the other end of the tether is connected with a metal wire to a recording device that ultimately is connected to a computer. The research subject has complete freedom of movement on the horizontal axis, but cannot move up and down.

Monarchs are sensitive to ultraviolet light, a wavelength of sunlight that is invisible to the human eye, and can use it to detect the sun's angle even on a cloudy day. When the flight simulator was covered with an ultraviolet filter, video monitoring showed that each of the Monarchs stopped flying within 15 seconds of applying the filter. Scientists have known for years that this ability is somehow involved in allowing them to navigate. Reppert and others further investigated this phenomenon.

The flight simulator is also able to demonstrate the importance of the Monarch's internal clock to their ability to navigate. Butterflies kept in the laboratory under a day/night cycle (light from 7 A.M. until 7 P.M.) consistent with the fall were taken outdoors in the flight simulator. They flew in a southwesterly direction as they would have naturally flown from the area where they were captured. Some Monarchs were kept under light conditions

The Monarch Man
Frederick Albert Urquhart, 1911–2002

Fred Urquhart was born in Toronto. He collected butterflies as a child, and this interest led him to complete his Ph.D. in entomology in 1940. He became an internationally renowned expert in the migration patterns of the Monarch butterfly. Dr. Urquhart was appointed Curator of Insects at the Royal Ontario Museum, was cross-appointed to the Department of Zoology at the University of Toronto, became Director of Zoology and Palaeontology at the Royal Ontario Museum, and in 1961 was appointed Professor of Zoology, a position he held until his retirement in 1979.

Urquhart and Norah R. Patterson were married in 1945 and began a lifelong project, rearing thousands of Monarch butterflies in order to study them and track their autumn migration. Urquhart had begun this work in 1937, and together they continued to tag thousands of Monarchs, affixing a tiny label to the wing of each butterfly which read, "Send to Zoology University of Toronto Canada." Tagged specimens were returned, and the researchers recorded the distances and locations where the marked individuals were found. Over the next 20 years, thousands of volunteers assisted with the tagging and participated in the project, yielding a wealth of information. The locations where the butterflies were recaptured indicated that they had migrated from the northeast in a southwesterly direction, leading the Urquharts to search for their overwintering habitat from the Gulf of Mexico to California.

Figure 20A. A close-up of one of the tags that thousands of volunteers apply to thousands of migrating Monarchs so that the butterflies' paths and destinations can be tracked.

Local people living in Mexico became involved in the search, and in 1975 the Urquharts got a call reporting that millions of Monarchs had been located in a relatively *(continued)*

The Monarch Man (*continued*)

small area in the oyamel fir forest, about 240 kilometers from Mexico City. The area is in the volcanic highlands of central Mexico, at an elevation of almost two miles. In 1976 the Urquharts visited the site, and it is said that a branch fell to their feet from the weight of the butterflies, and one of the butterflies had one of their tags on its wing that had been applied in Minnesota.

A dozen or so sites on five mountains in Mexico have been identified as winter habitat for Monarchs, and these areas are now protected as ecological preserves by the Mexican government, largely through Urquhart's early influence and advocacy. In Canada attention is focused on maintaining milkweed, the sole food of Monarch larvae, another of Uruqhart's concerns. A public butterfly garden named in their honor was established in Dundas, Ontario, in 1994, and in 1998 the Urquharts were appointed Members of the Order of Canada to honor them for their research and for their advocacy of protecting the Monarch's habitat.

that shifted their sense of time by six hours backward, and some by six hours forward. When taken outdoors in the flight simulator, the clock-shifted butterflies oriented 90 degrees to the west or the east of the direction they should have flown. Butterflies placed in the flight simulator after being housed under constant light flew directly toward the sun, presumably having lost all sense of time. This research provides strong evidence that migratory Monarchs use an internal time-compensated sun compass. New research by Zhu suggests a genetic basis to this ability.

Research focusing on the use the earth's magnetic field suggests Monarchs do not rely on magnetic orientation as a means of navigation. Monarchs subjected to a magnetic pulse when released flew in random directions, presumably not able to orient themselves. This demonstrates that the butterflies are sensitive to magnetic fields, but they do not necessarily use them for orientation. Indoor experiments in the same flight simulator used to prove the use of the sun compass showed that migratory Monarchs flew in all directions under normal, amagnetic, and reversed magnetic conditions. Some other cues, perhaps large topographical features, such as mountain ranges and bodies of water, may help the Monarchs find their way.

Question 5: Do all migrating Monarchs go to the same place?

Answer: North American Monarchs migrate each winter to areas with more favorable conditions (see Chapter 8, Question 2: Why do they migrate?). They do not travel to destinations where high temperatures would cause them to be constantly active and deplete the fat reserves needed for breeding and beginning the trip north in the spring. Instead, they go to cool, moist locations and enter a state of torpor, where they are semi-dormant. Here they can wait for the regrowth of their host plant, milkweed, across their breeding range.

Every year, approximately 100 million Monarchs, find their way to one of a dozen or so sites in a relatively small area (155 square kilometers) of Oyamel fir forests in the volcanic highlands of central Mexico's mountains. At almost two miles (3,000 meters) above sea level, the southwest-facing slopes are close to but not quite freezing. Fog and clouds provide water for the millions of clustered butterflies, while the branches of the Oyamel fir trees act as blankets and umbrellas, giving protection from the wind, rain, and occasional snow. Remaining clustered together and dry, the Monarchs can even withstand short periods of freezing temperatures.

A smaller number of Monarchs travel to groves along the coast of southern California. Here they congregate by the tens of thousands in eucalyptus, Monterey pine, and Monterey cypress trees, sheltered from harsh weather.

It has long been thought that the Rocky Mountains separate eastern and western Monarchs, with the western population migrating to the California coast and the eastern population migrating to Mexico. However, as observed by Pyle, at least some Monarchs west of the Rockies migrate to Mexico. Only the western most Monarchs overwinter on the California coast.

Question 6: How long does it take Monarch butterflies to migrate south?

Answer: Every fall, millions of Monarch butterflies migrate south from various points in eastern North America. Those emerging from their pupae in Canada, at the northernmost part of their range, can fly over 2,500 miles (4,023 kilometers).

The time taken to make the journey is dependent on weather conditions, since Monarchs do not fly when it is too cool (below 55 degrees Fahrenheit), when it is raining, or when a strong wind is blowing in the wrong direction. Usually the flight south takes two to three months. Much is known about the time and length of the Monarch's migration because, since it is so predictable, many organizations and researchers have gathered data annually on the phenomenon (see Chapter 8, Question 3: How do scientists study migration patterns? and Chapter 8, Question 10: How far can butterflies fly without stopping to rest?).

Question 7: What allows the migrating generation of Monarchs to live so long?

Answer: On average, the generation of Monarchs that emerges in the fall, prepared to migrate, lives eight or nine months. The three or four so-called summer generations that emerge during the rest of the year live only two to six weeks. In human terms, given an average lifespan of 75 years, this would be like having children who lived to be about 600 years old.

The individuals that emerge in the north in late August and September are physically prepared to migrate, and so they are somewhat different from the individuals that emerge in the summer. Studies by Goehring suggest some environmental cue, perhaps shortening day length or cooler nights, causes the caterpillar or pupae to suspend the full development of the

Circadian Rhythms

If you have ever traveled a long distance by airplane to a different time zone, you have probably experienced jet lag, with its associated symptoms of fatigue and insomnia. You felt out of sorts because your sleep/wake cycle was disrupted. Your internal biological clock is your *circadian rhythm,* and these rhythms are important in determining the sleeping and feeding patterns of all animals.

The term *circadian* comes from the Latin words *circa,* meaning around or approximately, and the word *dia,* which means day. Thus it literally means *approximately a day,* referring to the roughly 24-hour light/dark cycle in the physiological processes of living beings. There are clear patterns of brain wave activity, hormone production, cell regeneration, and other biological activities linked to this daily cycle. The level of the hormone melatonin rises in your body during the night and falls during the day, which is what causes jet lag to occur. Many aspects of animal behavior are governed by these rhythms.

Eclosion, the emergence of an insect larva from a pupa, is an example of butterfly and moth behavior that is governed by a biological clock. Most pupae of the same species tend to eclose at the same time of day, often at first light in butterflies, or after dark in many moth species.

Egyptian Cottonworm moths (*Spodoptera littoralis*) only mate during certain hours. The coordinated behavior in this and other species is a result of the synchronized release of sex pheromones that stimulate their mating rituals. Since the search for a mate is restricted to specific times of the day or night when mating is most likely to occur, the potential for successful reproduction is enhanced.

The fall generation of Monarch butterflies depends on the light/dark cycle to accomplish its long annual migration. They use a time-compensated sun compass to navigate to their overwintering grounds in Mexico. Experiments found

(continued)

Circadian Rhythms (*continued*)

that when their circadian rhythms were disrupted, the direction of their flight was changed.

Circadian rhythms tend to persist for a while, even when animals are kept in total darkness in the laboratory. But the biological clock can be reset, and most species adapt over time to changes in the day/night cycle. When you have traveled to another time zone, after a few days you probably have adjusted so that you slept and awoke at the appropriate times, indicating that your biological clock had been reset.

reproductive organs, and the adult emerges in a state of reproductive diapause. Though the butterfly looks the same as a summer Monarch, it is not sexually mature. These individuals are not able to reproduce in four to six days like summer Monarchs, but must wait until warmer weather triggers them to develop to maturity.

The inability to mate and reproduce extends the lifespan of the migrating Monarch, but they also need to prepare to survive the long journey and overwintering period ahead. They feed well on carbohydrate-rich nectar and convert the energy into fat tissue, known as the *fatbody,* in their abdomen. Research by Taylor, at the University of Kansas, shows the fatbody is an important reserve, and when the butterflies are unable to feed, breakdown of the fats produces the energy and water they need to survive.

They stop to nectar during the course of the migration in order to further sustain themselves, and a butterfly may gain or lose mass depending on whether it is able to feed or if it is using up stored energy. Monarchs can vary in size depending on sex, genetics, and nutrition, and butterflies of the same size can differ considerably in mass due to differences in the size of the fatbody. The mass of the fatbody the butterfly accumulates may be an important factor in its success in completing the migration. Scientists are studying how it is possible for such a small organism to travel so far and how they conserve their energy during flight (see Chapter 8, Question 10: How far can butterflies fly without stopping to rest?).

Question 8: How do Monarchs know when to migrate?

Answer: When the late summer and early fall monarchs emerge from their pupae in reproductive diapause, environmental factors indicate the onset of unfavorable conditions and the butterflies cannot reproduce and must head south (see Chapter 8, Question 7: What allows the migrating generation of Monarchs to live so long?). A series of experiments by Betalden and others at the University of Minnesota shows that the factors that trigger their instinctive migration are a combination of decreasing day length, cooler night temperatures than day temperatures, and decreasing host plant quality.

The situation is similar in anticipation of the arrival of spring. The Monarchs begin to leave Mexico and start to travel northward when the milkweed plants begin to sprout in March in the southern states near the Gulf of Mexico. The Monarchs follow the new growth of milkweed as the warm weather spreads northward, laying their eggs on the plants as they travel north.

Question 9: How do migratory Monarchs know where to go?

Answer: Since the Monarchs migrating to the over-wintering grounds have never been there before, we know it must be instinctive—or genetically programmed—rather than a learned behavior. Environmental cues tell them when to go (see Chapter 8, Question 8: How do Monarchs know when to migrate?), and we have some understanding of how they find their way (see Chapter 8, Question 3: How do scientists study migration patterns? and Chapter 8, Question 4: How do Monarchs navigate over long distances?), but how they know exactly where they need to go remains a mystery.

Question 10: How far can butterflies fly without stopping to rest?

Answer: No one knows for sure how far a butterfly can travel before stopping to rest, but studies by tagging programs have recorded data on how far individuals travel in a day (see Chapter 8, Question 3: How do scientists study migration patterns?). The record flight from the Urquhart tagging program was 265 miles (426 kilometers) in a single day. After many years of gathering information, it is generally thought Monarchs average between 40 to 100 miles per day.

In order to make the long journey as efficiently as possible, Monarchs take advantage of favorable wind directions and they glide to conserve energy. Soaring higher than 10,000 feet (3,000 meters), the butterflies ride on air currents that speed them southward. Research has shown that gliding flight is essential in order for Monarchs to complete the long journey. Gibo estimated that a Monarch can glide for 1,000 hours on the same amount of energy that would be consumed in only 44 hours if it were actively flying.

Question 11: Does each butterfly travel south to Mexico and back to the United States or Canada?

Answer: The journey south is long and hazardous, and many butterflies do not even make it to the overwintering grounds. However, those that survive the winter mate in the early spring and begin to head north in search of milkweed. In March the butterflies move into the southern states, the females laying eggs as they go. A rare few individuals may return further north, but the vast majority die out along the way, and it is their descendants that continue the northern flow of Monarchs.

NINE

Outdoor Butterflies

Question 1: Are butterflies and moths of any ecological value?

Answer: Butterflies are of great ecological importance for many reasons: they are pollinators, they are part of the *food web* (see below), and they are a good measure of the health of their habitats (*ecological indicators*).

About 80 percent of all flowering plants depend on insects to transfer pollen from one blossom to another. The pollen is carried from flower to flower on the insect's legs as it feeds on nectar from the blossoms. Many night blooming plants depend on moths for pollination, and butterflies and moths are second only to bees as pollinators. The process is vital to the production of food, since without pollination, plants would be unable to bear fruit or berries.

The term *food web* describes the feeding relationships among species in an ecosystem. Caterpillars and adult butterflies and moths are an essential part of the food web. In addition to eating leaves on living plants, many caterpillars eat fallen leaves, fruit, or decaying wood or animal matter. This helps break down and recycle nutrients in the habitat. In turn, caterpillars and butterflies are a food source for a wide range of predators, including other insects, tiny parasites, rodents, birds, frogs, and even humans (see Chapter 7, Question 1: Do people eat butterflies and moths?).

Butterflies are very sensitive to environmental change, which makes them excellent indicators of ecosystem health. Their disappearance from an area often signals an imbalance in the local system as a result of habitat destruction, pollution, or the overuse of pesticides (see Chapter 9, Question 3: Why don't I see as many butterflies as I used to?).

Question 2: Is it true that some butterflies and moths have a negative impact on the environment?

Answer: Some species of butterflies and moths are so well adapted to an environment and reproduce so well that their populations grow out of control. At their worst, the caterpillars eat so voraciously that their munching can be heard.

Probably one of the most well-known North American pests is the Gypsy moth (*Lymantria dispar*). Native to Europe, it was introduced to the United States in 1869 by Leopold Trouvelot, who had hoped to crossbreed it with U.S. silkworms. While cultivating the moths in his garden in Massachusetts, some larvae escaped into the nearby woods. Knowing the caterpillars were destructive, he immediately tried to find them with the help of others, but unfortunately the escapees were not found.

About 20 years after the accidental introduction, the area began experiencing such serious outbreaks of Gypsy moth populations that the state and federal government began making attempts to eradicate the moth. The Gypsy moth's preferred host plant is oak, but it also readily feeds on hundreds of species of trees and shrubs, including aspen, apple, hawthorn, sweetgum, and conifers. Its range has extended throughout the northeastern United States, south to Virginia, and west into Michigan and Wisconsin, with sporadic outbreaks as far west as California. The U.S. Forest Service reports that since 1980, the Gypsy moth has defoliated about 1 million acres (4,000 square kilometers) of forest each year.

Apple crops across North America are seriously affected by a number of moth pests, including the Codling moth (*Cydia pomonella*), the Redbanded Leafroller (*Argyrotaenia velutinana*), the Apple Ermine moth (*Yponomeuta malinellus*), and the Apple

Pith moth (*Blastodacna atra*). The larvae of these species eat the foliage, the fruit, or both.

The Codling moth larva is the well-known "worm in the apple," but it also attacks pears, crab apples, English and black walnuts, quince, and other fruits. Like the Gypsy moth, it was also introduced from Europe.

Until the 1950s the Redbanded Leafroller was only a minor problem, but following the introduction of DDT, an organic pesticide, it became a major pest. The DDT did not affect the moths themselves, but it did kill off the parasites that had kept the moth populations in check (see Chapter 7, Question 2: What dangers do caterpillars face?).

Pests also plague banana crops. The Banana Skipper (*Erionota thrax*) primarily eats bananas, although other recorded host plants include bamboo, Manila hemp, coconut, and other palms. Davis and Kawamura recorded the species as a notorious pest of crops in South and East Asia, Indonesia, the Philippine Islands, and Guam.

There are many other examples of moths and butterflies that are considered agricultural pests. Attempts to control them include the use of chemical or biological pesticides as well as introducing known predators or parasites. The Animal and Plant Health Inspection Service (APHIS), a branch of the U.S. Department of Agriculture, is responsible for protecting America's animal and plant resources from pests and diseases. They work together with other agencies to monitor and control the spread of established pests and to help prevent the introduction of potential new ones.

Question 3: Why don't I see as many butterflies as I used to?

Answer: Change is a normal and constant process, and butterfly populations wax and wane naturally over periods of many years. But because butterflies are so dependent on the continual existence of a very specific habitat to maintain a healthy population and good breeding conditions, they are vulnerable to even the slightest imbalance (see Chapter 6, Question 4: What do caterpillars eat?). Human interference in the environment causes stress on the ecosystem, and nature often cannot adapt quickly enough to compensate for the changes in the habitat.

Some of the major threats to ecosystems are habitat loss due to industrial and urban development, intensive farming practices, pollution, and overuse of pesticides. Intense fire suppression, which would seem to be a means of protecting habitats, can actually lead to profound ecological changes. Rudolph reported that in the Ouachita Mountains of Arkansas, decades of suppressing fires that would have naturally occurred has caused the forests to be dominated by hickory and oaks, making the canopy much denser and allowing less light to reach the forest floor. This restricts the growth of nectar sources and host plants and has led to a decline in the population of the Diana Fritillary (*Speyeria diana*) and the Great Spangled Fritillary (*Speyeria cybele*). The introduction of invasive exotic species is also a problem, since they may be predators or may breed more successfully and out-compete native butterflies (see Chapter 9, Question 7: Why is it important to know the difference between native and exotic species of plants and animals?).

Draining wetlands and reclaiming heathland for agricultural use has led to seriously declining numbers of the Marsh Fritillary (*Euphydryas aurinia*) and the Silver-studded Blue (*Plebejus argus*) in the United Kingdom. Poorly planned housing, industrial development, and construction of new roads can cut through and fragment critical habitats. Fragmentation of habitats, where pockets of suitable breeding grounds are left, like islands in a sea of development, is a serious threat to butterfly survival. Lack of connecting areas leaves the populations in each fragment isolated and unable to intermingle and breed.

Global warming is causing new climate conditions and patterns that in turn are leading to changes in growing seasons, patterns of seasonal breeding, and as reported by Parmesan, a shift in species ranges (the area where a species can normally live and breed). Longer, hotter summers and/or shorter winters can interfere with both the growth of host plants and the breeding cycle of butterflies and moths. Successive drought in an area eventually leads to changes in the predominant vegetation, which causes changes in the ecosystem. The climactic changes also result in an increase in natural disasters such as hurricanes and tropical storms. Recovery from the devastation caused by these events takes many years, and many species simply cannot adapt and survive.

Question 4: How can I encourage butterflies to visit and breed in my garden?

Answer: Whether you have a large garden, a small backyard area, or planter boxes on a balcony, with a little research and planning it is possible to attract butterflies. There are many books and websites dedicated to information on butterfly gardening, including information relevant to specific regions.

There are a few general notes to keep in mind when planning your butterfly garden. Try to include both flowering nectar plants for the adult butterflies and host plants for the caterpillars (see (see Chapter 9, Question 5: What is a nectar plant? and Chapter 9, Question 6: What is a host plant?).

If you only provide nectar sources, butterflies may visit to feed, but they will move on, looking for suitable breeding sites. By intermingling host plants with nectar plants, you will have the opportunity to witness the entire life cycle of the butterfly. Butterflies prefer to visit flowers in sunny, open patches, so avoid planting taller species where they cast shadows across shorter blooming plants. Include a variety of plants with different colored blooms, since some butterflies may be more attracted to one color over another.

It should be easy to find out what butterfly species are local to your area from a regional field guide or from your local chapter of the North American Butterfly Association (see Appendix E). Then you can provide nectar and host plants preferred by the species you wish to attract.

Question 5: What is a nectar plant?

Answer: Butterflies and moths visit flowers to sip the energy-rich nectar they produce, which contains water, small amounts of proteins, lipids, antioxidants, minerals, vitamins, and usually about 15–25 percent sugar. Many flowering plants are considered good nectar producers, though the amount and concentration of nectar produced can vary with the weather and time of day. Cold winds may decrease nectar production, while warm, sunny weather increases abundance and concentration. Nectar output is often highest around midday. Also, different plant species are better suited to different geographical regions. Climate, soil type, and other factors may contribute to the quality of nectar produced by a flowering plant, which will usually perform better in the region to which it is native (see Chapter 9, Question 7: Why is it important to know the difference between native and exotic species of plants and animals?).

Most butterflies will feed from a variety of flowering plants, but they often have a preferred nectar source. The Monarch (*Danaus plexippus*), for example, prefers milkweed, aster, and zinnia. To provide nectar sources in your garden from the spring through to the fall, look for plants that bloom at different times of the year.

For a few suggestions to get started, see Appendix A on nectar plants. Regional books and websites on butterfly gardening contain zone maps to help you choose plants suitable to your geographical area.

Question 6: What is a host plant?

Answer: Although caterpillars are eating machines, they tend to be selective about the leaves they eat, usually choosing a single plant species or a group of closely related species on which to feed. Some species are *polyphagous,* and will sustain themselves on a wider variety of plant species. The plants that caterpillars eat are known as host plants.

For example, Monarch butterflies (*Danaus plexippus*) eat milkweed (*Asclepias* spp.). Question Mark butterflies (*Polygonia interrogationis*) are less specific, eating elm and hackberry, and occasionally common nettle and hops. The Summer Azure's (*Celastrina neglecta*) preferred host plant varies with the season; early broods host on dogwood and cherry, later broods eat sumac, viburnum, and some legumes. The diet of the Io moth (*Automeris io*) varies geographically, and includes hackberry, cherry, maple, willow, clover, and birch.

Including host plants of local butterfly species in your garden will give you the opportunity to witness all stages of the life cycle. It can also be rewarding to include host plants for moths, especially the attractive giant silkworm moths (Saturniidae), such as the Luna (*Actias luna*), Polyphemus (*Antheraea polyphemus*), or Cecropia (*Hyalophora cecropia*) moths. Host plants include a huge variety of types, from herbaceous flowers to shrubs and trees, and many host plants are what people consider weeds. Converting at least part of your garden to a wilder, overgrown, native habitat will benefit butterflies and attract other insects and birds, too. For host plant suggestions see Appendix B.

Caterpillars are voracious eaters and will strip your plants of foliage while they are feeding and growing. The plants may not look pretty, but you will be rewarded with a bounty of butterflies and moths as the caterpillars develop.

Question 7: Why is it important to know the difference between native and exotic species of plants and animals?

Answer: Native plants and animals are ones that have evolved together in a region over thousands of years and so are uniquely adapted to local conditions. Exotic species are those that have been introduced, either accidentally or intentionally, to an area where they were not previously found. It is helpful to be aware of which species are not native to your locality, because over time some exotic species may become invasive.

When considering which plants to choose for a butterfly garden, in the interest of conserving native habitats remember that your local butterflies have evolved alongside the local plants, so you will need host plants that are native species if you want the butterflies to breed. Native plants flourish with less attention, needing no fertilizer and little watering, since they are suited to the local environment.

Good places to find out more about plants native to your area are regional wildflower guides, nature centers and botanical gardens, local gardening clubs, and your local chapter of either Wild Ones or the North American Native Plant Society (see Appendix E). Also visit local nurseries and farmers markets, and an Internet search will provide numerous companies from which to order supplies.

Question 8: What other garden features can I provide for butterflies?

Answer: The right plants will provide nectar for butterflies and food for caterpillars. However, if you have the space you may want to consider providing for some other needs that butterflies have, such as basking, drinking, and puddling (see Chapter 3, Question 6: What are butterflies doing when they gather on the ground?). Meeting these needs will encourage butterflies to stay longer in your garden.

Rocks or logs piled in a sunny area will invite butterflies to perch and bask when they are not feeding, and they will provide nooks in which to shelter during inclement weather. A shallow birdbath or dish can be a good water source for butterflies, and gravel or pebbles placed in the water will allow them to stand and drink without getting their legs and wings wet.

A shallow container filled with muddy water, or just a patch of damp, bare earth can be a good spot for butterflies to puddle and obtain nutrients and essential salts. Puddling is especially important to males, who lose minerals and salts in the sperm packets (spermatophore) they transfer to females during mating.

If you are able to tolerate a certain amount of "untidiness" in your garden, you can attract butterflies that prefer a more unusual menu. Allowing fallen fruit to rot and ferment will attract a number of species, such as the Question Mark (*Polygonia interrogationis*) and Mourning Cloak (*Nymphalis antiopa*), but it can also attract wasps. Oozing sap from broken branches or insect- damaged bark is attractive to the Red Admiral (*Vanessa atalanta*), while animal dung and carrion will be visited by the Eastern Comma (*Polygonia comma*) and the Red-spotted Purple (*Limenitis arthemis astyanax*).

Question 9: Can I use pesticides in my butterfly garden?

Answer: If you have a problem with unwelcome insects on your plants, such as aphids, scale, or mealybugs, try using natural pesticides or biological controls. Chemical pesticides and herbicides will be harmful to the butterflies and caterpillars you are trying to attract. If the infestation of plant pests is not severe, a first approach is to try treating the plant by spraying the leaves at close range with a high-pressure mister on a water hose. The impact of the water washes off the offending insects with no lasting damage to the leaves, or to caterpillars that may later feed on them. You might also consider pruning infested stems or entirely removing infested plants.

Commercially available insecticidal soaps or even homemade soapy water are effective at controlling plant pests. Crenshaw advises that soaps and detergents work as contact insecticides, with a 2 to 3 percent solution applied directly onto the insects, coating them thoroughly. Homemade solutions using household soaps are less expensive than commercial preparations, but they might cause plant injury (*phytotoxicity*). Since the soap is only effective when wet, wash the treated leaves a few hours after the application to reduce the risk of injury to the foliage.

In general, insecticidal soaps are effective against most small, soft-bodied arthropods, such as aphids, young scales, whiteflies, mealybugs, and spider mites. Soap sprays are considered selective insecticides because of their minimal effects on other organisms, such as caterpillars and beneficial insects such as pollinating bees and ladybugs.

Commercial soaps are usually sodium or potassium salts combined with fatty acids, but some contain other insecticides, such as pyrethrins or neem, so check the label to be sure the soap you choose does not contain anything potentially harmful. Although it is a natural product derived from the neem tree seeds, neem is harmful to caterpillars as well as to the soft-bodied insects you are trying to eliminate. Just because products are labeled organic or "natural" does not necessarily mean they are nontoxic to beneficial insects, pets, or even humans

Gardeners who are interested in protecting the environment have increasingly found it helpful to use biological controls. For example, the ladybug species *Harmonia axyridis* and *Hippodamia convergens* prey on aphids, with a single ladybug beetle consuming as many as 50 aphids a day. Many companies sell small colonies of a wide array of beneficial insects, such as whitefly and aphid parasites, predatory mites, bees, and ladybug beetles. *It is important to be aware that most gardening companies selling biological controls also consider caterpillars to be pests, so make sure the colony you are buying will not harm caterpillars.*

For more information on specific plant pests and on natural methods to control them, an Internet search using the key words "natural pesticides" or "biological pest control" will yield copious information. You will also find links to many companies supplying products and biological controls.

Question 10: Do people still collect butterflies?

Answer: Collecting butterflies for your own enjoyment, either for a preserved collection or for breeding at home, is fine for the vast majority of species. Care should be taken to collect responsibly, being mindful not to damage any habitat and taking only a limited number of common specimens from a location. Conscientious collecting remains a vital tool of butterfly conservation, due to the atlas and distributional studies it makes possible. It is impossible to conserve organsims without a clear and precise picture of where they occur. Collecting is usually prohibited in parks and nature reserves, and butterfly species that are severely threatened are legally protected from being harmed or collected. (See Chapter 9, Question 11: Are any species of butterflies threatened or endangered?)

During the 16th and 17th centuries large numbers of butterflies were taken back to the old world by explorers to be described, classified, and studied by scientists. During the 19th century, collecting butterflies became fashionable for wealthy individuals and for scholarly naturalists (see the Carl Linnaeus sidebar on page 37). The demand for butterflies for collections and study led to the development of butterfly farms, which raised and shipped the livestock in the United Kingdom and the United States and from tropical areas around the world.

As a result of the interest of many serious collectors, over 2 million butterflies and moths were eventually donated to the Museum of Natural History in London, and important 18th-century collections were donated to other museums as well, such as the Natural History Museum (Naturalis) in Leiden, The Netherlands (see photographs in the insert from the Naturalis collection, donated by Pieter Cramer in the late 18th century).

Starting a collection is not difficult, and the techniques for preparing specimens can easily be mastered and the minimum equipment required is not expensive. A starter kit, containing all you need, may be obtained from scientific supply companies, such as BioQuip (see Appendix D). At the very least you will need a net, small envelopes or vials in which to transport specimens home, small spade-tip forceps, a spreading board, good quality insect pins, labels, and a protective box in which to store your collection.

A butterfly or moth may be frozen for 24–48 hours, in order to kill it humanely. If it is not possible to prepare the specimen soon after collection, it can be stored in a glassine or paper envelope in a dry environment and relaxed and spread at a later date. Specimens prepared on a spreading board will need to dry out thoroughly for three to ten days, depending on size, before being transferred to your storage box.

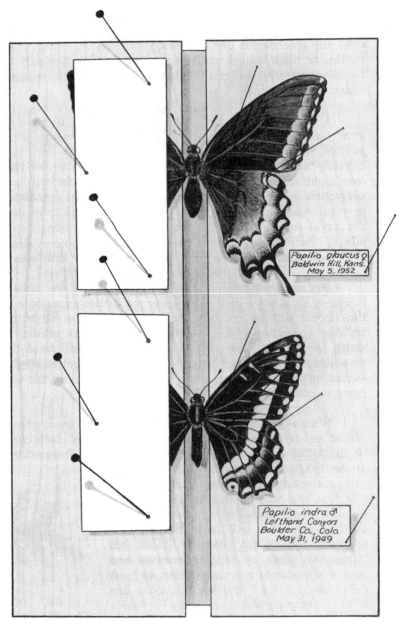

Figure 22. A mounting board showing spread butterflies. Paper strips (left) keep wings in place until dry. *(Drawing courtesy of William H. Howe, from* Our Butterflies and Moths, *page 24)*

To be of scientific value, all specimens in a collection should be carefully labeled. The first label should include the precise location of the collection site, how the specimen was collected, the date of capture, and the name of the collector. The second label should list the scientific name of the specimen.

Most collectors concentrate on adults, but it is possible to preserve other life stages as well. To preserve larvae, bring a pan of distilled water to a boil, remove it from the heat, then drop in the specimens for a few seconds. Then transfer the specimens into a vial of 70 to 80 percent ethanol solution for storage.

More people these days are turning to butterfly collections in the form of photographs or keeping lists of butterfly sightings. Collecting images of butterflies and moths in the wild can yield a lot of information and is less invasive on butterfly populations. (See Chapter 9, Question 16: Are there any tips for photographing butterflies?)

Question 11: Are any species of butterflies threatened or endangered?

Answer: As a species's population becomes more and more reduced, it may be considered vulnerable, threatened, or endangered. If a species is reduced to a very small number of individuals it is in danger of becoming extinct (see Chapter 9, Question 12: Are any butterfly species extinct?). Butterflies are sensitive to environmental changes, and in many areas around the world numerous species have greatly declined in population due to a variety of factors (see Chapter 9, Question 3: Why don't I see as many butterflies as I used to?).

There are international, national, and local wildlife agencies that strive to protect threatened and endangered species, and which impose stiff penalties for illegal collecting, harassing, or habitat disturbance. Founded in 1948, the International Union for Conservation of Nature and Natural Resources (IUCN), also known as the World Conservation Union, assembles and publishes conservation objectives for all at-risk plants and animals, through its *Red Data Books*. Many countries have adopted the valuable IUCN guides and maintain their own Red Lists of vulnerable, threatened, and endangered species, including butterflies.

Worldwide recognition and listing of endangered species led to the 1975 Convention on International Trade in Endangered Species of Wild Fauna and Flora (CITES). Agreements between governments ensure that trade in listed specimens is restricted and does not threaten the survival of imperiled species. CITES lists roughly 5,000 species of animals, including 67 species of Lepidoptera, all belonging to the Swallowtail family (Papilionidae). The Queen Alexandra's Birdwing (*Ornithoptera alexandrae*), the world's largest butterfly, is now so rare in the wild that it is an endangered species. Loss of habitat has been a major factor in its decline, but this beautiful butterfly has been highly prized by collectors for many years and over-collecting of individuals has contributed to its scarcity. It is now protected from being collected and traded under international laws by CITES. A critically endangered butterfly may be protected under both international and national law. The largest true swallowtail in the Americas, the Jamaican Giant Swallowtail (*Papilio homerus*) was listed by CITES in 1987 and the Jamaican Wildlife Act in 1988.

In the United States, butterflies may be protected by listing with a federal agency or by an individual state. The U.S. Fish and Wildlife Service lists over 20 species, including the Karner Blue (*Lycaeides melissa samuelis*), the Oregon Silverspot (*Speyeria zerene hippolyta*), and the Schaus Swallowtail (*Heraclides aristodemus ponceanus*). The Endangered Species Act of 1973 allows for federal agencies to impose fines and imprisonment under criminal and civil codes. All of the protected species are listed on the USFWS website, www.fws.gov. Some butterflies in the United Sates, not covered by federal listing, are covered under individual state listing, such as the rare Miami Blue (*Hemiargus thomasi bethunebakeri*). When the species was added to Florida's endangered species list in 2003, stiff fines of $5,000 and imprisonment afforded the butterfly protection.

Another organization, the Xerces Society, which is dedicated to protecting all imperiled insects, maintains the Red List of Pollinator Insects of North America. Listed are over 55 species of at-risk butterflies and moths found in Mexico, the United States, and Canada, including the Blackburn's Sphinx moth (*Manduca blackburni*), found only in Hawaii (see Chapter 9, Question 13: What are people doing to protect butterflies?).

In the United Kingdom about half of the 56 resident butterfly species have seriously reduced populations. Several of the species, including the High Brown Fritillary (*Argynnis adippe*) and the Heath Fritillary (*Melitaea athalia*) are at a critical level due to habitat destruction.

Question 12: Are any butterfly species extinct?

Answer: If a species is no longer able to sustain a population and has died out it is considered extinct. Perhaps the most well-known American butterfly to become extinct is the Xerces Blue (*Glaucopsyche xerces*). It was once locally common on the San Francisco Peninsula, but was last seen there in March 1941. Extinction of the species was probably due to urban development from the growth of San Francisco.

A species may be considered *possibly extinct* if over a long period of time no individuals are seen in an area where it was once found. Although the Lotis Blue (*Lycaeides argyrognomon lotis*) was never a common butterfly, known only from a few areas on California's north coast, it was recorded consistently for over 100 years. The last population was observed in Mendocino County, California, in 1994. Surveys of the Lotis Blue's last known location, conducted in 2000, did not find any of the butterflies or its host plant, seaside bird's-foot trefoil. In some instances, a species thought possibly extinct will be rediscovered as an isolated population or after a natural disturbance, such as a hurricane.

A species may completely disappear from an area where it was once common, yet it may still be found in other places. These species become known as *extirpated* or *locally extinct*. Since 1864, five species of butterflies have become locally extinct in the United Kingdom, most recently the Large Blue (*Maculinea arion*) in 1979 and the Large Tortoiseshell (*Nymphalis polychloros*) in the 1980s. All five of the species, though extinct in the United Kingdom, have remained common to various places in continental Europe. In 1983 a project to reintroduce the Large Blue to the United Kingdom was established using Swedish caterpillars to rebuild the population (see Chapter 9, Question 13: What are people doing to protect butterflies?).

Question 13: What are people doing to protect butterflies?

Answer: Many local and international organizations are dedicated to protecting butterflies and their habitats. The scale of the programs they administer varies, but all focus on the need to halt the growing number of vulnerable, threatened, and endangered species and to protect the habitats those species depend upon. The organizations welcome public involvement, and it is likely you can find a butterfly conservation project in your area.

The new California Academy of Sciences building in Golden Gate Park in San Francisco will have a "green" roof, planted with California native plants. Green roofs have many ecological benefits, and it is hoped that this one will attract the endangered San Bruno Elfin (*Callophrys mossii bayensis*).

The Butterfly Conservation Initiative was established in 2001, through the partnership of six organizations, with the objectives of supporting the recovery of federally listed endangered species and increasing public awareness and involvement in local conservation efforts across the United States. The coalition develops and supports programs for endangered, threatened, and vulnerable species, such as the Karner Blue (*Lycaeides melissa samuelis*) and the Oregon Silverspot (*Speyeria zerene hippolyta*).

The recovery plans are tailored to the conservation needs of each species and target appropriate action. For some species, working with local landowners to protect habitat may be most important, for others captive rearing and reintroduction programs help re-establish dwindling populations. Additional approaches to aid recovery are raising and planting host plants, protecting existing habitat by encouraging native plant species and removing exotics, and raising public awareness of particular species.

Historically the Karner Blue occurred in 12 states throughout the Northeast and Midwest and at several sites in the province of Ontario, Canada. However, due to habitat loss and fragmentation of remaining suitable habitat, populations were reduced to only part of its former range. The recovery plan focuses on management of existing habitat and creating new habitats. Local programs are run in association with several zoos and have encouraged schoolchildren of all ages to get involved with growing and planting wild lupine (*Lupinus perennis*), the Karner Blue's host plant. Also, through a breeding program, reintroductions are underway at three sites, in New Hampshire, Indiana, and Ohio. (See Chapter 9, Question 11: Are any species of butterflies threatened or endangered?)

The North American Butterfly Association (NABA) is the largest nonprofit, butterfly organization in the United Sates, Canada, and Mexico. Its aim is to increase the public's enjoyment of butterflies and, in conjunction with other organizations, help with the conservation of butterfly species, in particular Florida's Miami Blue (*Hemiargus thomasi bethunebakeri*).

Miami Blue butterflies were common throughout coastal southern Florida 50 years ago, but by the late 1980s their population had declined steeply due to habitat fragmentation (see Chapter 9, Question 3: Why don't I see as many butterflies as I used to?). Hurricane Andrew in 1992 devastated the last strongholds of the butterfly, and it was feared the species had become extinct. Fortunately, in 1999 a tiny colony of perhaps only 50 individuals was discovered in Bahia Honda State Park, near Key West.

It was clear immediate action was needed to conserve the vulnerable Miami Blue. Although the butterfly was not granted federal protection, NABA petitioned the Florida Fish and Wildlife Conservation Commission to have it added to the state list of endangered species. The listing ensured legal protection for the Miami Blue (see Chapter 9, Question 11: Are any species of butterflies threatened or endangered?), and conservation efforts shifted to research and recovery. In 2003 a breeding program lead by Jaret Daniels was launched by the University of Florida's McGuire Center for Lepidoptera and Biodiversity with wild collected eggs, and the first reintroductions of captive raised adults and caterpillars began in early 2004. By 2006, approximately 30 generations later, more than 20,000 individuals have been bred in captivity, and some 3,800 released into suitable locations. Continued field research and monitoring programs aid the Miami Blue's recovery.

Officially listing a species can be a sensitive issue in areas undergoing land development, since the new legal status of a threatened or endangered species can affect zoning laws. When Eric Quinter discovered a potentially new and rare species of butterfly on North Carolina's Crystal Coast, nicknamed the Crystal Skipper (*Atrytonopsis* new species 1) it was exciting for scientists, but a potential headache for developers and private landowners. Listing the species is under debate because of the possible impact on the rapidly developing barrier island called Bogue Banks.

Founded in the United Kingdom in 1968, Butterfly Conservation aims "to conserve butterflies and moths and the habitats on which they depend." The society works to halt the decline of UK butterflies and moths in many ways, including conserving and restoring important habitat, creating and managing nature reserves, advising landowners, devising and managing species action plans for threatened and endangered species, and encouraging greater public awareness of butterflies and moths.

In partnership with several other organizations, the society has been involved with the Large Blue Project. The project aims to rebuild the population of the Large Blue (*Maculinea arion*), which had been locally extinct in the United Kingdom since 1979, probably due to changes in farming practices (see Chapter 9, Question 12: Are any butterfly species extinct?).

Research discovered that the Large Blue needed short grass and warm soil, the preferred habitat of the *Myrmica sabuleti* red ant. Like many other species in the Lycaenidae family, ants tend the caterpillars (see Chapter 6, Question 4: What do caterpillars eat?). In the case of the Large Blue, Jeremy Thomas discovered its caterpillars would only survive in a nest of that particular ant species. As part of the project, grazing was re-established on suitable sites and the remaining colonies of ants quickly spread. The species host plant, thyme, was planted and in 1983 caterpillars were released. Their habitat has been carefully managed and an estimated 10,000 Large Blue butterflies were recorded by the summer of 2006.

Question 14: Where can I go butterfly watching?

Answer: Butterfly watching, or *butterflying*, is now as popular a pastime as bird watching. And butterflies are most active when they are warm, so it is not necessary to be out in the field at the crack of dawn to observe them; in fact, early in the day they are likely to be asleep unless it is very warm. Special equipment is not necessary, and if you are patient, it is possible to get quite close to resting or nectaring butterflies so you can observe and photograph them. If you are interested in using binoculars, close-focusing binoculars make it possible to focus clearly on a subject only a few feet away, and this type of binocular is better for butterfly watching than ones you might use for birding. Some enthusiasts like to have a net and a hand lens for temporary capture and observation.

When looking for butterflies among flowers, also be alert for wet areas, rotting fruit, animal carcasses, and animal droppings—all places where butterflies are likely to be found. And if you are searching with a group, be conscious of butterflying etiquette. Everyone should stop when a butterfly is spotted, and you should be sure everyone gets a look before you quietly inch forward for a closer look or for a photograph. Try not to startle the butterfly with your shadow or with a sudden movement, and back away carefully when you are done so that someone else may have a turn. Be careful not to trample any foliage as you move around.

States That Have an Official State Butterfly

State	Butterfly
Alabama	Eastern Tiger Swallowtail and Monarch
Arizona	Two-tailed Swallowtail
Arkansas	Diana Fritillary
California	California Dogface
Colorado	Colorado Hairstreak
Delaware	Eastern Tiger Swallowtail
Florida	Zebra Longwing
Georgia	Eastern Tiger Swallowtail
Idaho	Monarch
Illinois	Monarch
Kentucky	Viceroy
Maryland	Baltimore Checkerspot
Minnesota	Monarch
Mississippi	Spicebush Swallowtail
Montana	Mourning Cloak
New Hampshire	Karner Blue
Oklahoma	Black Swallowtail
Oregon	Oregon Swallowtail
South Carolina	Eastern Tiger Swallowtail
Tennessee	Zebra Swallowtail
Texas	Monarch
Vermont	Monarch
Virginia	Eastern Tiger Swallowtail
West Virginia	Monarch

Close to home your local botanical garden or nature center may have a dedicated butterfly garden or wild meadow area to attract butterflies. City, state, and national parks provide important habitats, and may have checklists of the local butterfly species. Contact your area chapter of the North American Butterfly Association (NABA) or the Lepidopterists' Society to find out about guided walks or field trips you can join (see Appendix E).

Local museums and colleges may also offer informative outings. There are over 200 public conservatories around the world where you can observe butterflies in a naturalistic setting (see Chapter 10, Question 1: What is a butterfly conservatory?). Appendix C lists some of these conservatories and their contact information.

You may be interested in volunteering to help with a tagging program, which monitors Monarch migration. The Monarch Program, Monarch Watch, and Journey North are organizations that need volunteers to help with observations. The annual "Fourth of July Butterfly Count," initiated by the Xerces Society and coordinated by NABA, welcomes all butterfly enthusiasts. Counts are held in multiple locations, so contact NABA to find one in your area. International organizations, such as the Earthwatch Institute, provide opportunities for volunteers to help scientists with their field projects.

The growth in ecotourism means more travel companies are including wildlife watching as part of their tours, especially to Central and South America. Many specialized companies tailor the entire itinerary to experiencing the flora and fauna of a habitat, with the benefit of specialist naturalists leading the tours.

Question 15: How can I see more moths?

Answer: A good place to start *mothing* is at existing light sources. These may be porch lights, walls of buildings next to well-lit parking lots, highway rest stops, or storefronts. The illuminated restrooms and shower facilities at campgrounds are often full of moths drawn to the bright light.

If you want to observe moths in a more remote location, contact your local Lepidoptera group or wildlife center to find out about organized mothing in your area. You can also set up your own light source near a large white bed sheet to attract moths. The bed sheet can be laid out flat on the ground, draped over the hood of your car, or hung from a line between posts. Different species of moths fly at different times of night. Some

Figure 23. A mercury vapor lamp and sheet set up at night to attract moths for observation and collection. *(Photograph by Hazel Davies)*

fly soon after dark, whereas others are not on the wing until the early hours of the morning. You may want to experiment with lighting at different times.

Lights designed specifically for attracting moths are available from scientific supply companies, such as BioQuip (see Appendix D). The two main types are mercury vapor (MV) and fluorescent "blacklights." Blacklights emit the shorter ultraviolet wavelengths attractive to moths, but they are less powerful. MV lights have excellent ultraviolet output and overall greater intensity. MV bulbs get very hot and can shatter if rain or mist rolls in. They are best used with a dedicated rain shield or an up-turned Pyrex bowl.

The process of "sugaring," attracting moths to a sugar mixture painted onto plants or rocks, was used as long ago as 1841 reports Young. Recipes vary, and some lepidopterists guard the secret of their ingredients closely, but the basic ingredients are molasses or any unrefined sugar, stale beer, and rotten fruit of any kind. The mixture should ferment for several days before use, and some enthusiasts recommend adding a shot of dark rum.

A Bird in the Bush Is Worth Two in the Hand: The Origins of Butterfly Watching

Before the turn of the 20th century, people interested in butterflies or birds killed and collected them in order to learn about them. Entire flocks of birds were slaughtered for their plumage. John James Audubon (1785–1851), the famous painter of birds in naturalistic poses and settings, is said to have believed that it was not a good day if he shot fewer than 100 birds, and all of his paintings were of carefully arranged dead birds.

The creation of the Audubon Society by George Bird Grinnell in 1886 marked the beginning of the U.S. conservation ethic. Although there are references to binoculars going back as far as 1608, the first high-quality modern binoculars were sold in 1894, and this gave a great impetus to bird watching as a replacement for the shooting of birds. By the early 20th century, more than 15 state Audubon Societies had been formed, and they were focusing interest on protecting wildlife and their habitats.

Butterfly collectors were still active, but the idea of observing butterflies without catching and killing them was an outgrowth of the conservation movement and became increasingly popular and politically correct. Robert Michael Pyle's field guide, *Watching Washington Butterflies*, published in 1974, was among the first books that popularized butterfly watching, and it was followed by many more butterfly identification books, distribution maps, reports on migration patterns, handbooks, checklists, books about planting butterfly gardens, and books of personal anecdotes about the pleasures of observing butterflies and moths. At the present time, there are many interest groups, clubs, societies, newsletters, journals, field trips, and festivals devoted to butterflies. Some of the popular ways to pursue an interest in butterflies are:

Observing butterfly behavior at close range in the field.

Participating in nighttime moth trips.

(continued)

A Bird in the Bush (*continued*)

Visiting butterfly conservatories.

Collecting caterpillars and/or pupae and observing the entire life cycle.

Learning to identify different species and keeping a checklist of what you have seen.

Planting a butterfly garden.

Participating in conservation projects and local butterfly counts.

Volunteering to participate in a tagging program to study migration patterns.

Once ready, the mixture can be painted onto tree trunks, fence posts, and rocks along a route you can repeatedly check over a few hours, such as along the edge of woodland. Apply the bait at dark and wait about a half hour before your first excursion along the route. Sugaring repeatedly over several nights can improve results, but tends to be most successful on calm, damp nights. It will also attract butterflies during the day.

Winter describes a method useful in treeless situations known as "wine-roping." Take a boiled and cleaned cotton clothesline and soak it in a bait mixture comprised of 2 quarts (2 liters) of cheap red wine and 1 pound (half a kilogram) of brown sugar. The rope can be draped along hedgerows or fences and has a similar effect to sugaring.

Question 16: Are there any tips for photographing butterflies?

Answer: Most modern digital cameras allow you to capture acceptable or good photographs, especially if you are shooting in a well-lit conservatory where butterflies are plentiful. Locating and approaching the butterflies you wish to photograph outdoors takes a little practice and patience, but there are a few useful things to remember that will make it easier for you to approach a butterfly without scaring it away.

Trying to photograph a butterfly often requires that the photographer assume a very awkward position in order to frame the shot from a pleasing angle. A camera with autofocus makes it possible to shoot with one hand, leaving the photographer a spare hand to brace or balance with or to hold a reflector to angle more light on the subject.

If you are interested in consistently recording great photographs of the butterflies you see outdoors, you will probably want to consider investing in some minimal equipment, which will greatly enhance your shots. To capture detail and obtain a good, frame-filling shot you will need either a 35mm single lens reflex camera or the best high-resolution digital camera you can afford. Since most butterflies are small, to get frame-filling pictures, choose a camera with a true macro lens, which will result in a life-size image.

Getting close to your subject with a macro lens means having to address the issue of lower light levels for focusing. Many photographers prefer to use natural light, but the combination of the increased magnification and a moving subject may not allow for a long exposure. It may be necessary to use a flash to help freeze the motion of the butterfly, counteract the effects of a breeze, or even minimize blur caused by your own breathing. However, if you are very close to the subject the flash mounted on the top of your camera may be too powerful, causing the butterfly to be washed out. Instead, you might opt for a bracket-mounted flash or a ring flash, depending on your camera and budget. A simple reflector made from a circle of shiny silver fabric stretched over a flexible hoop is a fairly inexpensive accessory, available from most camera stores. It can really improve a shot by bouncing natural light onto the dimly illuminated underside of your subject.

An image that is life-size or larger tends to pose focusing problems. It narrows the depth of field to a fraction of an inch, meaning that even a small butterfly may not be in focus from one wing tip to another. To improve the focus, get parallel with your subject, looking directly from behind or above the butterfly as the situation allows. Or shoot a resting butterfly at a 90-degree angle to the folded wings.

Butterflies prefer warm, sunny conditions and you will generally find them most active during the middle part of the day. Visit local nature preserves and state parks, or look for concentrations of flowering plants alongside trails or disused railroads or even in empty parking lots. In order not to startle the butterfly, approach slowly enough so that the butterfly does not respond to the movement, be aware of your shadow so as not to allow it to pass over and frighten the butterfly, but move quickly enough to compose your photograph before the butterfly moves on.

When you see a butterfly nectaring, it may seem flighty at first, but you will notice that it stops to feed at intervals. If it flies away be patient, as it will probably return. This may happen several times until eventually it seems to surrender and just feed and ignore you.

Butterflies you find puddling (see Chapter 3, Question 6: What are butterflies doing when they gather on the ground?) may stay still for a relatively long time. If the day is sunny but cool, you may find a butterfly basking to raise its body temperature so it can fly. A basking butterfly usually will stay quite still, hoping to go unnoticed, and in that state you can often photograph them to your heart's content. Butterflies feeding on rotten fermented fruit can often become quite inebriated and sluggish, making them easy to approach and photograph.

As you gain experience, keep in mind a few tips to help improve the composition of your shot. Try to move around the butterfly to shoot from different angles. A clear photograph of open wings from above is great for identification, but also try getting down to eye level for a more intimate image. Is the background distracting? If so, try to position a leaf behind the butterfly or use the narrow depth of field to blur the background.

When moving in for the perfect shot, be careful not to trample the nectar plants on which the butterfly is feeding. If you are with a group of photographers or butterfly watchers, Glassberg suggests the proper etiquette is to let all members of the group

get a good look at the subject and to wait for them to take photographs from a distance before you approach the butterfly to
try to get your close-up shots.

Question 17: Is it safe to release butterflies at weddings and other events?

Answer: There is much debate over the popular trend of releasing commercially raised butterflies at special events. Some
scientists maintain that the releases pose a threat to wild populations, and some observers ponder whether it is morally wrong
to exploit living organisms for people's amusement. Commercial butterfly breeders maintain that they only sell healthy butterflies, and that there is no evidence that the releases have a
negative impact on the environment.

The intentional release of native birds was outlawed over
50 years ago, and NABA president Glassberg and other Lepidopterists now think it is time to do the same with butterflies.
In the United States, butterfly releases are regulated by the
Department of Agriculture, which only allows permits for nine
species. Just two species, the Monarch (*Danaus plexippus*) and
Painted Lady (*Vanessa cardui*), constitute about 99 percent of the
releases. Those calling for the ban on butterfly releases cite past
examples where accidental or intentional releases of animals
have produced negative results (see Chapter 9, Question 2: Is it
true that some butterflies and moths have a negative impact on
the environment?), noting that the full impact of the practice
may not be seen until many years into the future.

Scientists worry about the potential spread of parasites and
diseases among butterfly populations. This is of special concern in Monarchs, which can be infected with the debilitating
protozoan parasite *Ophryocystis elektroscirrha* in both wild and
farm-raised populations. The International Butterfly Breeders
Association maintains that members take extreme measures
to protect their stock and "do not release or use for breeding
any livestock that indicates the prescence of disease" (see Chapter 3, Question 13: Do butterflies carry diseases?).

Often butterflies are released a great distance from where they were raised, so the potential to introduce butterflies that do not normally occur in an area is very high. Glassberg notes this could confuse species distribution and migration studies, with sightings recording a species outside of its normal range, or may result in inappropriate genetic mixing of widely separated species. For example, in Monarch populations it is thought this could lead to problems with migration because the ability to navigate might be compromised. Other scientists and suppliers contend that there is no evidence that the genetic make-up of the natural populations would be altered.

The breeders association note that many of the statements made against releases are based around things that "might or could" happen, not hard facts, and that some opponents are letting their emotions take over because they just "do not want to see butterflies commercialized." But raising butterflies is a business. Releasing butterflies at events has grown in popularity, and it is estimated that butterfly farms in the United States ship out millions of butterflies a year. With some species costing up to $10 each, NABA worries that the commercial market created for live butterflies has led to poaching of wild individuals, especially at Monarch overwintering sites (see Chapter 8, Question 5: Do all Monarchs go to the same place?). Whatever the arguments for and against releases, it is likely the debate will continue until monitoring and research prove what, if any, impact is being made on wild populations.

TEN

Indoor Butterflies

Question 1: What is a butterfly conservatory?

Answer: A butterfly conservatory is an enclosed garden stocked with free-flying butterflies and sometimes moths. A conservatory may bring together many species from around the world or it may consist of only local species that are bred in the conservatory. Most butterflies in conservatories are purchased from butterfly farms and shipped as pupae (see Chapter 10, Question 5: Where do the butterflies come from?). The butterflies and moths emerge soon after their arrival and are released into the conservatory.

A conservatory may be located completely indoors, or it may be located in a glass greenhouse or a screened-in outdoor area. Conservatories, or butterfly exhibits as they are also known, are often found on the grounds of a zoo, botanical garden, or other public space. Walking paths are provided so that visitors can circulate through the exhibit while they observe and photograph the various species with minimal disturbance to the butterfly's natural behavior. Some conservatories are elaborately landscaped and decorated with waterfalls and exotic plants, while others are casually planted or are simple greenhouses.

Indoor conservatories often exhibit exotic tropical species, so they are artificially heated to mimic the natural climate of the butterflies. Outdoor gardens in temperate regions are often open for only part of the year, since butterflies and moths flourish in warmth, and local weather conditions must be suitable.

Question 2: When did conservatories first appear?

Answer: Rudimentary conservatories were first developed over 2,000 years ago by the Romans as a way to protect and display delicate plants during cold weather. They constructed temporary wooden buildings that kept out cold but let in light, using thin sheets of semi-transparent mica (a silicate mineral) as glazing.

During the 16th century indoor gardens became fashionable with wealthy landowners in England, Holland, and France. These winter shelters gave protection from the weather, and made it possible to cultivate and study cuttings and new plant species brought back from maritime expeditions to the New World, Africa, and Asia. The early conservatories had panels that could be removed to let in more light on mild days, and candles, small fires, or stoves were used to provide heat. The cultivation of tropical fruit trees became very popular at this time, and the structures became known as *orangeries*. They were fashionable places for entertaining, and during summer months the plants were moved outdoors and the interiors were used for social gatherings as described by de Vleeschouwer.

Technological advances in the production of wrought and cast iron in the 19th century made it possible to build conservatories with more elaborate designs that spanned larger areas. Systems for providing heat, water, and ventilation were also improved. In the early 1800s the first public urban greenhouses were built in Paris and London. In the United States, the first public greenhouse was built for the centennial exhibition in Philadelphia in 1876. Cunningham calls them *crystal palaces*.

Beginning in the 19th century it became popular to enhance the plant displays in greenhouses with the addition of animals, particularly birds and butterflies. Collecting butterflies became a hobby of "nature aristocrats," who had the time and money to indulge. The first live butterfly shipment to cross the Atlantic is said to have contained a Blue Morpho caterpillar (*Morpho rhetenor*) and enough host plants to survive the journey, although it is not clear exactly when this occurred.

The demand for butterfly specimens led to the development of butterfly farms in Brazil in the mid-19th century. When live butterflies were first imported to England, they commonly were shipped in triangular paper envelopes. This mode of transportation resulted in a mortality rate often exceeding 40 percent. Professional butterfly farms that could provide a reliable source of butterflies began to develop and ship livestock from Asia, and later from Latin America, Oceania, and Africa. The first profitable butterfly farm in the United Kingdom opened in 1894, and one of the earliest butterfly farms in the United States opened in 1912 in California.

Worldwide Butterflies started in 1961 as a supply business for lepidopterists, and they invited visitors as early as 1962 to tour the breeding areas. A display of living Lepidoptera was opened to the public in 1973, but it was not until 1976 that they introduced the idea of tropical plants with free-flying butterflies inside a brightly lit glass structure that people looked in upon. It was a huge success, and led to the opening in 1978 of the Palm House at Compton House, where visitors could walk among the free-flying butterflies.

The increased ease of shipping pupae by air with express carrier services made it possible to safely display exotic butterflies from around the world, and this increased the popularity of conservatories. The first major permanent butterfly exhibit in North America was Insect World at the Cincinnati Zoo, which opened in 1978. The Calgary Zoo housed the first Canadian live butterfly display in 1986, and in 1988 Florida's Butterfly World opened its doors.

Exhibits have proliferated all over the world, ranging from simple seasonal greenhouse displays to magnificent year-round facilities with extravagant plantings, waterfalls, and extensive educational components. There are approximately 200 exhibits in operation, some temporary, some seasonal, and some permanent. (See Appendix C for a list of some of the conservatories, and check your local media for new or temporary seasonal exhibits.)

Question 3: Why are conservatories often so hot?

Answer: The exotic species exhibited in many conservatories are usually from the tropics, since the tropics have such a huge diversity (see Chapter 4, Question 8: Why are most butterflies found in the tropics?). The butterflies need to be provided with an environment as close as possible to their native climate, so entering a conservatory often feels like you have stepped into a hot and humid rainforest. The ideal conditions are a temperature of approximately 80 degrees Fahrenheit (27 degrees Celsius) and about 80 percent relative humidity.

Question 4: Do butterflies reproduce in the conservatory?

Answer: Many conservatories in the United Sates displaying exotic species do not allow the butterflies to reproduce, usually due to the strict regulations governing butterfly exhibits. Species from other countries can only be imported into the United States with a permit issued by the Department of Agriculture, and the permit may not allow breeding of imports.

Though butterflies will mate in a conservatory, females will only lay eggs on specific host plants (see Chapter 9, Question 6: What is a host plant?). Therefore, where breeding must be controlled, conservatories use only nonhost plants in their gardens and landscaping so that the butterflies will not lay eggs and reproduce.

Different countries have different regulations governing importation and rearing; therefore, in some parts of the world even nonlocal species may be allowed to breed in conservatories. And, exhibits displaying local species may breed butterflies by providing host plants. In this type of conservatory the visitor is able to see all the life stages, and can easily spot caterpillars feeding on the foliage of their host plants.

Question 5: Where do the butterflies come from?

Answer: The butterflies in a conservatory may be bred locally, but they often come from distant butterfly farms. The majority of farms are located in tropical areas around the world where, due to the climate, it is possible to breed butterflies year round. Butterfly farming is popular in Florida, Central and South America, Africa, Southeast Asia, and Australia.

Some butterfly farms breed and maintain all their stock inside greenhouses. Other farms have outdoor mesh flight houses, where the adults mate and lay their eggs. The eggs are then harvested and raised in an indoor lab for protection from parasites and disease. Farmers cultivate a continual supply of host plants for the caterpillar species they breed, monitor the growing larvae, and ensure they all have enough food. The farmers keep a percentage of the pupae they raise to maintain their breeding stock.

A different method, known as *butterfly ranching,* uses wild adults. Gardens rich in nectar sources and host plants are developed on the edge of existing forest, encouraging the adults to visit, mate, and lay eggs. Ranchers monitor the developing larvae by either placing nylon mesh bags over the plants to contain the caterpillars and afford them some protection from predators, or by moving them to wooden framed, screened rearing cages. Butterfly ranching projects are proving to be effective ways to conserve habitat while providing income to local people. Earning more from farming butterflies than they did raising other crops, local ranchers become protectors of the environment by preventing deforestation and clearing of native habitats.

Farmers and ranchers harvest fresh pupae daily and prepare them for the journey to conservatories by carefully wrapping the chrysalides in tissue or cotton wool and packing them securely in a sturdy box. The butterflies are transported in the pupal stage, since that is the easiest life stage to pack and ship. Accompanied by export and import permits for customs checks and agricultural inspections, the pupae may travel three or four days to international destinations. The butterflies emerge from the pupae at the conservatory, usually within a week of arrival.

Question 6: Do all the different species get along?

Answer: All the species coexist peacefully in a conservatory, even though they might not encounter one another in their native habitats. If any mild aggression does occur, it is usually due to competition between males of the same species. In some species the males patrol a territory and chase away other males that enter the space in order to keep competitors away from the local females (see Chapter 7, Question 12: Do butterflies fight?).

Figure 24.   Staff at Costa Rica Entomological Supply meticulously inspecting pupae for parasites or disease before packing and shipping them to butterfly exhibits. *(Photograph by Hazel Davies)*

Figure 25. Several fruit-feeding species, including Owl butterflies (*Caligo* spp.), feast on ripe mango and banana in a conservatory. (*Photograph by Hazel Davies*)

Question 7: What do you feed the butterflies?

Answer: There are usually many varieties of flowering plants in conservatories, which provide nectar for the butterflies (see Chapter 3, Question 3: What do butterflies eat? and Chapter 9, Question 5: What is a nectar plant?). Popular nectar-producing plants include pentas, ixora, jatropha, and porterweed. A highlight of visiting a conservatory is seeing a butterfly fluttering from one flower to the next, inserting its proboscis to drink nectar. Conservatories often also provide dishes of artificial nectar, a mixture of either sugar or honey and water, to make sure there are ample food sources available for all the butterflies.

Some butterflies prefer to feed on fruit, so dishes of bananas, oranges, melons, or mangos are provided for these species. Fruit-eating butterflies include Owls (*Caligo* spp.) and Blue Morphos (*Morpho* spp.) from South and Central America. These butterflies cannot pierce the fruit with their soft proboscis, but they drink the sugar-rich juice from the ripe fruit.

Question 8: What happens at night?

Answer: The environment in a conservatory is designed to closely simulate the natural environment of the butterflies, even when indoors. At outdoor exhibits and glass conservatories the butterflies become less active during the evening as the light naturally fades. The butterflies respond by looking for a place to rest and spend the night, usually hanging from the underside of a leaf or from a branch. In exhibits with artificial lighting, at the end of the day the lights dim gradually, to indicate that darkness is approaching, and the lights are turned completely out at night to provide a day/night cycle (see Chapter 3, Question 9: Do butterflies sleep?).

Some butterflies are *crepuscular* (most active at dawn and dusk), so it is important for these species to have a gradual change from light to dark in the exhibit, when they may spend some time feeding. Moths in the conservatory usually become active at night, since most species are nocturnal and they have spent the day resting.

Question 9: Do indoor butterflies recognize their keepers?

Answer: Butterflies do not see in the same way we do and there is no evidence that they can recognize people. On average butterflies live only 10 to 14 days, during which time they are busy feeding and looking for mates.

Question 10: How can I attract a butterfly to land on me?

Answer: Conservatory butterflies are still wild animals. If a butterfly does not land on you when you visit a conservatory it does not mean the butterflies do not like you. Butterflies in conservatories, just like those in the wild, spend their time looking for food, water, basking spots, and mates.

A butterfly may land on you if it thinks you are a flower, possibly because your perfume or the scent of your hair product attracts them. Or it may land on your head or hand if you are sweating, so that it can drink the moisture and salts—often bald men's heads are popular landing sites!

Some conservatories allow you to spray water on your hand to attract the butterflies to land and drink. However, not all do this, so you should always respect the rules in the conservatory you are visiting. A butterfly may land on you by chance, but if it does not, you can still enjoy observing their fascinating behavior close-up.

Question 11: Can I raise butterflies at home?

Answer: Rearing butterflies at home or in the classroom is not complicated or costly and can be a very educational and rewarding experience. When carefully observed, the caterpillars in your care will yield a great deal of information about their behavior and development. Raising butterflies provides the amateur entomologist with a wonderful opportunity to document and photograph all stages of a species' life cycle.

Caterpillars can be collected individually by carefully searching foliage. When you find one, pick off the leaf and pop it into a secure container. A more efficient method to collect caterpillars from trees and shrubs is using a beating sheet or simply a white bed sheet. Expert Wagner notes that caterpillars can hold on tenaciously if disturbed, so you need to catch them by surprise. Smoothly bend a branch over your collecting sheet and then strike it firmly to dislodge the caterpillars from the leaves. Carefully remove any leaves or twigs from your sheet and inspect for caterpillars. A sweep net can be used to collect grass-feeding species. Take care when handling caterpillars, as some can have irritating hairs or spines that may be allergenic to some people (see Chapter 7, Question 3: How do caterpillars defend themselves?).

Small plastic containers make ideal homes for individual caterpillars, while glass jars or larger food storage boxes are adequate for a small colony. You can pierce very small holes in the lid, or cut out a window and cover it securely with fine screen mesh to allow for ventilation. Alternatively, you can set up mesh cages or place mesh bags over host plants in your garden.

Larger colonies are prone to the outbreak of disease, and overcrowding can lead to smaller adults. For an optimal environment provide fresh foliage daily and keep the containers clean and free from mold and condensation. Placing paper towels in the base of the container soaks up excess moisture and makes removing frass easier. If your caterpillar is a species that pupates in the ground, you need to provide a layer of slightly damp peat in the base of the container during the final instar (see Chapter 6, Question 6: How does a caterpillar grow?). Provide sticks ready for the newly emerged adult to climb on and hang from while it dries its wings.

Butterfly kits, most commonly containing Painted Lady butterflies (*Vanessa cardui*), are readily available from scientific and educational supply companies. The larvae received with the kit can be raised to adulthood in the same way as you might raise locally collected caterpillars. However, the question of what to do with the resulting butterflies is quite different. The adults from locally collected caterpillars can easily be released into the wild in the same location from which they came. Larvae sent in a kit have probably been bred a great distance from where you will raise them, and releasing them into the wild could have serious consequences for the local butterfly population (see Chapter 9, Question 17: Is it safe to release butterflies at weddings and other events?). It is far better to keep mail-order butterflies contained at home or in the classroom for the duration of their life and to learn not only about metamorphosis, but also about how it is important to protect the environment.

Selected Nectar Plants for North American Butterflies and Moths

Organized alphabetically by common name, the following list includes a selection of plants that are attractive to butterflies as a nectar source, and where indicated, as larval food. Where spp. follows the genus, two or more species within the genus are known to be attractive nectar sources. Also indicated are the type of plant, the growth height, and the bloom season. This list only represents a small fraction of the plants suitable for a butterfly garden. Please refer to regional gardening guides for more information on plants suitable to your geographical area.

* Indicates that the species is also listed as a host plant in Appendix B

*Alfalfa (*Medicago sativa*). Perennial; up to 2 feet; late spring to early fall.

*Aster (*Aster* spp.). Hardy perennial; 2 to 3 feet; late summer to fall.

Bee Balm (*Monarda didyma*). Hardy perennial; 2 to 4 feet; summer to fall.

*Black Cherry (*Prunus serotina*). Medium-sized tree; up to 100 feet; late spring.

Black-eyed Susan (*Rudbeckia* spp.). Annual that seeds itself; up to 2 feet; midsummer to fall.

Chives (*Allium schoenoprasum*). Perennial herb; 1 foot; summer.

*Chokecherry (*Prunus virginiana*). Small tree or shrub; up to 15 feet; late spring.

Coneflower, Purple (*Echinacea purpurea*). Perennial; 2 to 4 feet; midsummer to fall.

Cosmos (*Cosmos* spp.). Annual; 4 to 10 feet; late summer to fall.

*Dill (*Anethum graveolens*). Annual herb; 3 to 4 feet; midsummer.

Gayfeather (*Liatris* spp.). Hardy perennial; 2 to 3 feet; midsummer to fall.

Goldenrod (*Solidago rugosa*). Hardy perennial; 1 to 3 feet; late summer to early fall.

Heliotrope (*Heliotropium arborescens*). Perennial; 2 to 4 feet; late spring to summer.

Indian Paintbrush (*Castilleja coccinea*). Annual, biennial; 1 to 2 feet; late spring.

Ironweed (*Vernonia noveboracensis*). Perennial; 3 to 7 feet; late summer to fall.

Joe-pye Weed (*Eupatorium* spp.). Hardy perennial; 2 to 3 feet; late summer to fall.

Lantana (*Lantana* spp.). Annual in the north, perennial shrub or vine in the south; up to 6 feet; summer in the north, year round in warm southern climate.

Lavender (*Lavandula angustifolia*). Small shrub; 1 to 3 feet; summer.

*Lilac (*Syringa* spp.). Small tree or shrub; up to 10 feet; spring.

*Milkweed, Common (*Asclepias syriaca*). Hardy perennial; up to 4 feet; summer.

*Milkweed, Orange (*Asclepias tuberosa*). Hardy perennial; 1 to 3 feet; summer.

*Milkweed, Swamp (*Asclepias incarnata*). Hardy perennial; up to 4 feet; summer.

Mint (*Mentha* spp.). Perennial herb; 1 to 4 feet; summer to fall.

Oxeye Daisy (*Chrysanthemum leucanthemum*). Perennial; 1 to 3 feet; summer.

*Parsley (*Petroselinum crispum*). Biennial herb; 1 to 2 feet; midsummer to early fall.

Pentas (*Pentas lanceolata*). Annual in the north, perennial in the south; 1 to 3 feet; summer in the north, year round in warm southern climate.

Phlox (*Phlox* spp.). Hardy perennial; 2 to 4 feet; summer.

Pussytoes (*Antennaria* spp.). Perennial; less than 1 foot; early summer.

*Queen Anne's Lace (*Daucus carota*). Biennial; up to 3 feet; midsummer to early fall.

Rosemary (*Rosmarinus officinalis*). Small shrub; 2 to 6 feet; early spring.

Salvia (*Salvia lyrata*). Perennial; 1 to 2 feet; spring.

Shrubby Cinquefoil (*Potentilla fruticosa*). Small shrub; 2 to 3 feet; early spring to fall.

*Snapdragon (*Antirrhinum majus*). Annual; 1 to 2 feet; summer.

Sunflower (*Helianthus* spp.). Annual tall species, perennial smaller species; 3 to 10 feet; mid to late summer.

*Sweet Clover (*Melitotus* spp.). Perennial; 2 to 3 feet; late spring to fall.

Verbena (*Verbena* spp.). Annual, biennial, or perennial depending on species; up to 1 foot; spring to early fall.

*Viburnum (*Viburnum* spp.). Small tree or shrub; up to 15 feet; spring.

Whorled Stonecrop (*Sedum ternatum*). Perennial; less than 1 foot; late spring.

Yarrow (*Achillea* spp.). Perennial; 1 to 4 feet depending on species; late summer to fall.

Zinnia (*Zinnia* spp.). Annual; 1 to 3 feet; midsummer to fall.

Host Plants of Selected North American Butterflies and Moths

Swallowtails (Papilionidae)

Black Swallowtail (*Papilio polyxenes*)—Dill (*Anethum graveolens*), Parsley (*Petroselinum crispum*), and Queen Anne's Lace (*Daucus carota*)

Eastern Tiger Swallowtail (*Papilio glaucus*)—Ashes (*Fraxinus* spp.), Black Cherry (*Prunus serotina*), Spicebush (*Lindera benzoin*), and Tulip Tree (*Liriodendron tulipifera*)

Giant Swallowtail (*Papilio cresphontes*)—Citrus (*Citrus* spp.), Hop-tree (*Ptelea trifoliate*), and Torchwood (*Amyris elemifera*)

Spicebush Swallowtail (*Papilio troilus*)—Sassafras (*Sassafrass albidum*) and Spicebush (*Lindera benzoin*)

Whites and Yellows (Pieridae)

Cabbage White (*Pieris rapae*)—many plants in the mustard family (*Brassica* spp.)

Clouded Sulphur (*Colias philodice*)—Clover (*Trifolium* spp.) and other legumes

Orange Sulphur (*Colias eurytheme*)—Alfalfa (*Medicago sativa*), Clovers (*Trifolium* spp.), and Milk-vetches (*Astragalus* spp.)

Brushfoots (Nymphalidae)

Admiral, Red (*Vanessa atalanta*)—Nettles (*Urtica* spp.) and occasionally Hops (*Humulus* spp.)

Admiral, White (*Limenitis arthemis arthemis*)—Aspens (*Populus* spp.) and Birches (*Betula* spp.)

Common Buckeye (*Junonia coenia*)—Plantains (*Plantago* spp.), Snapdragon (*Antirrhinum majus*), and wild Petunias (*Ruellia* spp.)

Fritillary, Gulf (*Agraulis vanillae*)—Passion Flower Vine (*Passiflora* spp.)

Fritillary, Meadow (*Boloria bellona*)—Violets (*Viola* spp.)

Monarch (*Danaus plexippus*)—Milkweeds (*Asclepias* spp.)

Mourning Cloak (*Nymphalis antiopa*)—Aspens (*Populus* spp.), Elms (*Ulmus* spp.), and Willows (*Salix* spp.)

Painted Lady (*Vanessa cardui*)—wide variety of plants in many families, including Hollyhock (*Alcea rosea*), Mallow (*Malva* spp.), Nettles (*Urtica* spp.), and Thistle (*Cirsium* spp.)

Pearl Crescent (*Phyciodes tharos*)—Asters (*Aster* spp.)

Queen (*Danaus gilippus*)—Milkweeds (*Asclepias* spp.)

Question Mark (*Polygonia interrogationis*)—Elms (*Ulmus* spp.), Hackberries (*Celtis* spp.), Hops (*Humulus* spp.), and Nettles (*Urtica* spp.)

Red-spotted Purple (*Limenitis arthemis astyanax*)—Black Cherry (*Prunus serotina*), Chokecherry (*Prunus virginiana*), Oaks (*Quercus* spp.), and Poplars (*Populus* spp.)

Viceroy (*Limenitis archippus*)—Cottonwoods (*Populus* spp.) and Willows (*Salix* spp.)

Blues, Coppers, and Hairstreaks (Lycaenidae)

American Copper (*Lycaena phlaeas*)—Docks (*Rumex* spp.), especially Mountain Sorrel (*Oxyria digyna*) and Sheep Sorrel (*Rumex acetosella*)

Azure, Spring (*Celastrina ladon*)—Dogwoods (*Cornus* spp.), Privets (*Ligustrum* spp.), and Viburnums (*Viburnum* spp.)

Azure, Summer (*Celastrina neglecta*)—Cherries (*Prunus* spp.), Dogwoods (*Cornus* spp.), Sumacs (*Rhus* spp.), and Viburnums (*Viburnum* spp.)

Blue, Eastern Tailed (*Everes comyntas*)—many legumes, including Clovers (*Trifolium* spp.), Lupines (*Lupinus* spp.), and Sweet Clovers (*Melilotus* spp.)

Hairstreak, Banded (*Satyrium calanus*)—Hickories (*Carya* spp.), Oaks (*Quercus* spp.), and Walnuts (*Juglans* spp.)

Hairstreak, Coral (*Satyrium titus*)—Wild Plums (*Prunus* spp.)

Hairstreak, Great Purple (*Atlides halesus*)—Mistletoes (*Phoradendron* spp.)

Skippers (Hesperiidae)

Skipper, Dun (*Euphyes vestries*)—Sedges (*Carex* spp.).

Skipper, Fiery (*Hylephila phyleus*)—Grasses, especially Bermuda Grass (*Cynodon dactylon*) and Bluegrasses (*Poa* spp.).

Skipper, Sachem (*Atalopedes campestris*)—Grasses, especially Bermuda Grass (*Cynodon dactylon*).

Skipper, Silver-spotted (*Epargyreus clarus*)—Several legumes, including Locust (*Robinia* spp.) and Wisterias (*Wisteria* spp.).

Giant Silkworm and Royal Moths (Saturniidae)

Cecropia Moth (*Hyalophora cecropia*)—Apples (*Malus* spp.), Ashes (*Fraxinus* spp.), Birches (*Betula* spp.), Elms (*Ulmus* spp.), Lilacs (*Syringa* spp.), Poplars (*Populus* spp.), and Willows (*Salix* spp.).

Io Moth (*Automeris io*)—Huge variety of plants, shrubs, and trees, including Aspens (*Populus* spp.), Birches (*Betula* spp.), Cherries (*Prunus* spp.), Clovers (*Trifolium* spp.), Elms (*Ulmus* spp.), Oaks (*Quercus* spp.), Sassafrass (*Sassafrass albidum*), Willows (*Salix* spp.), and Wisterias (*Wisteria* spp.).

Luna Moth (*Actias luna*)—Many deciduous trees, including Birches (*Betula* spp.), Hickories (*Carya* spp.), Walnuts (*Juglans* spp.), and Willows (*Salix* spp.).

Polyphemus Moth (*Antheraea polyphemus*)—Many shrubs and trees, including, Apples (*Malus* spp.), Ashes (*Fraxinus* spp.), Birches (*Betula* spp.), Dogwoods (*Cornus* spp.), Elms (*Ulmus* spp.), Hazel (*Corylus avellana*), Hickories (*Carya* spp.), Rose (*Rosa* spp.), and Willows (*Salix* spp.).

Promethea Moth (*Callosamia promethea*)—Many shrubs and trees, including Ashes (*Fraxinus* spp.), Cherries (*Prunus* spp.), Lilacs (*Syringa* spp.), Sassafrass (*Sassafrass albidum*), Spicebush (*Lindera benzoin*), and Tulip Tree (*Liriodendron tulipifera*).

Royal Walnut Moth (*Citheronia regalis*)—Many plants and trees, including Ashes (*Fraxinus* spp.), Cherries (*Prunus* spp.), Hickories (*Carya* spp.), Lilacs (*Syringa* spp.), and Walnuts (*Juglans* spp.).

Public Butterfly Conservatories and Exhibits

In compiling this list, we wanted to show our readers that butterfly exhibits have opened all over the world. Many of the listed facilities are not open all year round, and some are temporary and may have closed by the time you read this book. It is necessary to contact any facility when you are planning a visit. We have included as much current contact information for each exhibit as we could find.

North America

Canada

Alberta

The Conservatory in the Dorothy Harvie Garden, Botanical Garden, and Prehistoric Park, Calgary Zoo
403-232-9300 / www.calgaryzoo.org

Butterfly House, Devonian Botanic Gardens—Edmonton
780-987-3054 / www.devonian.ualberta.ca

British Columbia

Butterfly World and Gardens—Coombs
250-248-7026 / www.nature-world.com/tropical.html

Victoria Butterfly Gardens—Brentwood Bay
250-652-3822 / www.butterflygardens.com

New Brunswick

Butterfly World—Moncton
506-852-9406 / www.butterflyworld.ca

Green Village Home and Garden Butterfly House—Fredericton
866-450-3388 / www.greenvillage.net

Newfoundland

Newfoundland Insectarium Butterfly Pavilion—Reidville
709-635-4545 / www.nfinsectarium.com

Nova Scotia

The Butterfly House at the Museum of Natural History—Halifax
902-424-7353 / www.museum.gov.ns.ca/mnh

Ontario

F. Jean MacLeod Butterfly Gallery, Science North—Sudbury
705-522-37.01 / www.sciencenorth.ca/science-attractions/butterfly-gallery/butterfly-gallery.html

Humber Nurseries Butterfly Conservatory—Brampton
905-794-0553 / www.gardencentre.com

Niagara Parks Butterfly Conservatory—Niagara Falls
905-356-8554 / www.niagaraparks.com/nature/butterfly.php

Urquhart Butterfly Garden—Dundas
905-627-5270 / www.unityserve.org/butterfly/

Wings of Paradise Butterfly Conservatory—Cambridge
519-653-1234 / www.wingsofparadise.com/

Quebec

L'Arche des Papillons—St. Bernard de Lacolle
450-246-2552 / www.larchedespapillons.com

Montreal Botanical Garden—Montreal
514-872-1400 / www2.ville.montreal.qc.ca/jardin/en/menu.htm

Papillons en Fete, Centre Jardin Hamel—Ancienne-Lorett
418-872-9705 / www.jardinhamel.com

United States

Alabama

Birmingham Zoo Butterfly Encounter—Birmingham
205-879-0409 / www.birminghamzoo.com

Tessman Butterfly House, Huntsville Botanical Center—Huntsville
256-830-4447 / www.hsvbg.org

Arizona

Maxine and Jonathan Marshall Butterfly Pavilion—Phoenix
480-941-1225 / www.desertbotanical.org

Butterfly Magic at Tuscon Botanical Gardens—Tuscon
520-326-9686 / www.tucsonbotanical.org

California

Butterflies Alive, Santa Barbara Museum of Natural History—Santa
Barbara
805-682-4711 / www.sbnature.org

Butterfly Garden at Six Flags Marine World—Vallejo
707-643-6722 / 707-644-4000 ext. 270 / www.sixflags
.com/discoverykingdom

Butterflies, Turtle Bay Exploration Park—Redding
800-887-8532 / 530-243-8850 / www.turtlebay.org

Butterfly Vivarium at Monarch Program—Encinitas
760-944-7113 / www.monarchprogram.org

Robinsons-May Pavilion of Wings, Natural History Museum—Los
Angeles
213-763-3466 / 213-763-3558 / www.nhm.org

Colorado

Butterfly Pavilion and Insect Center—Westminster
303-469-5441 / www.butterflies.org

Western Colorado Botanical Gardens and Butterfly House—Grand
Junction
970-245-3288 / www.wcbotanic.org

Delaware

Butterfly House, Ashland Nature Center—Hockessin
302-239-2334 / www.delawarenaturesociety.org

District of Columbia

Butterflies and Plants: Partners in Evolution, Smithsonian National
Museum of Natural History
202-357-1386 / www.mnh.si.edu

Florida

Butterfly Habitat, Gulfcoast Wonder and Imagination Zone—Sarasota
941-906-1851 / www.gwiz.org

Butterfly Rainforest at the McGuire Center for Lepidoptera and Biodiversity, Florida Museum of Natural History—Gainesville
352-846-2000 / www.flmnh.ufl.edu

Butterfly World at Tradewinds Park—Coconut Creek
954-977-4400 / www.butterflyworld.com

Greathouse Butterfly Farm—Earlton
866-475-2088 / www.greathousebutterflyfarm.com

Key West Butterfly and Nature Conservatory—Key West
305-296-2988 / www.keywestbutterfly.com

Kissimmee Good Samaritan Butterfly Garden—Kissimmee
800-859-1550 / www.goodsamkiss.com/butterfly.html

Lukas Butterfly Encounter—Oveido
407-365-6163 / www.lukasbutterflyencounter.com

Panhandle Butterfly House—Navarre
850-623-3868 / www.panhandlebutterflyhouse.org

Pollination Pavilion at Naples Botanical Garden—Naples
239-643-7275 / www.naplesgarden.org

Wings of Wonder Butterfly Conservatory, Cypress Gardens—Winter Haven
863-324-2111 / 941-324-2111 ext. 460 / www.cypressgardens.com

Georgia

Day Butterfly Center, Callaway Gardens—Pine Mountain
800-225-5292 / www.callawaygardens.com

Illinois

Butterflies! Brookfield Zoo—Brookfield
708-485-0263 / www.brookfieldzoo.org

Judy Istock Butterfly Haven, Peggy Notebaert Nature Museum, Chicago Academy of Sciences—Chicago
773-755-5100 / www.chias.org

Indiana

Butterflies and Bugs at the Botanical Conservatory—Fort Wayne
206-427-6440 / www.botanicalconservatory.org

Indianapolis Zoological Society, White River Gardens—Indianapolis
317-630-2001 / www.indyzoo.com

Iowa

Christina Reiman Butterfly Wing, Iowa State University—Ames
515-294-2710 / www.reimangardens.iastate.edu

Kansas

Botanica's Butterfly House, Wichita Gardens—Wichita
316-264-0448 / www.botanica.org

Louisiana

Audubon Insectarium—New Orleans (opening 2008)
504-861-2537 / www.auduboninstitute.org

Maryland

Wings of Fancy, Brookside Gardens—Wheaton
301-962-1400 / www.mc-mncppc.org/parks/brookside

Massachusets

Butterfly Garden at Museum of Science—Boston
617-723-2500 / www.mos.org

Butterfly Landing, Franklin Park Zoo—Boston
617-541-5466 / www.zoonewengland.org

Butterfly Place—Westford
978-392-0955 / www.butterflyplace-ma.com

Magic Wings Butterfly Conservatory—South Deerfield
413-665-2805 / www.magicwings.com

Michigan

Butterflies in Bloom, Dow Gardens—Midland
800-362-4874 / www.dowgardens.org

Butterfly Garden, Detroit Zoological Institute—Royal Oak
248-398-0900 / www.detroitzoo.org

Foremost's "butterflies are blooming," Lena Meijer Conservatory, Frederik Meijer Gardens and Sculpture Park—Grand Rapids
888-957-1580 / www.meijergardens.org

Mackinac Island Butterfly House—Mackinac Island
906-847-3972 / www.originalbutterflyhouse.com

Wings of Mackinac Butterfly Conservatory—Mackinac Island
906-847-3307 / www.wingsofmackinac.com

Minnesota

Butterfly Garden at the Minnesota Zoo—Apple Valley
800-366-7811 / www.mnzoo.org

Missouri

Butterfly Palace and Rainforest Adventure—Branson
417-332-2231 / www.thebutterflypalace.com

Monsanto Insectarium, Mary Ann Lee Butterfly Wing, St. Louis Zoo—St. Louis
800-966-8877 / www.stlzoo.org

Powell Gardens—Kingsville
816-697-2600 / www.powellgardens.org

Sophia M. Sachs Butterfly House, Missouri Botanical Garden—Chesterfield
636-530-0076 / www.butterflyhouse.org

Weinmeister House of Butterflies—Osage Beach
573-348-0088 / 800-461-6273 / www.houseofbutterflies.com

Nebraska

Butterfly Pavilion, Folsom Children's Zoo—Lincoln
402-475-6741 / www.lincolnzoo.org/butterfly_pavillion.html

New Jersey

Butterfly Exhibit, Bergen County Zoological Park—Paramus
201-262-3771 / www.fieldtrip.com/nj/12623771.htm

Kate Gorrie Memorial Butterfly House, Buttinger Nature Center—Pennington
609-737-3735 / www.thewatershed.org/education.php

Philadelphia Eagles Four Seasons Butterfly House, Camden Children's Garden—Camden
856-365-8733 / www.camdenchildrensgarden.org/butterfly.html

New Mexico

PNM Butterfly House, Albuquerque Biological Park—Albuquerque
505-764-6200 / www.pnm.com/community/butterfly.htm

New York

Butterfly and Moth Vivarium, Sweetbriar Nature Center—Smithtown
631-979-9233 / www.sweetbriarnc.org

Butterfly Conservatory, American Museum of Natural History—
New York
212-769-5100 / www.amnh.org

Butterfly House, Adirondack Visitors Center—Paul Smiths
518-327-3000 / http://www.adirondackvic.org/AboutVIC.html

Butterfly Zone, Bronx Zoo—New York
718-220-5100 / www.bronxzoo.org

Dancing Wings Butterfly Garden, Strong Museum—Rochester
585-263-2700 / www.strongmuseum.org

Stonehouse Farm Butterfly Conservatory—Oneonta
607-434-1506

Up Yonda Farm Environmental Education Center—Bolton Landing
518-644-9767 / www.upyondafarm.com

North Carolina

Magic Wings Butterfly House, North Carolina Museum of Life and
Science—Durham
919-220-5429 / www.ncmls.org

North Carolina Museum of Natural Sciences—Raleigh
919-733-7450 / www.naturalsciences.org

Ohio

Annual Butterfly Show, Krohn Conservatory—Cincinnati
513-421-5707 / www.butterflyshow.com

Butterfly House, Cox Arboretum—Dayton
937-434-9005 / www.coxarboretum.org/ButterflyHouse.htm

Butterfly House, Put-in-Bay—South Bass Island
419-285-2405 / www.perryscave.com

The Butterfly House—Whitehouse
419-877-2733 / www.butterfly-house.com

Cleveland Botanical Garden—Cleveland
216-721-1600 / www.cbgarden.org

Cleveland Metroparks Zoo—Cleveland
216-635-3335 / www.clemetzoo.com

Franklin Park Conservatory—Columbus
800-214-7275 / www.fpconservatory.org

Insect World, Cincinnati Zoo and Botanical Garden—Cincinnati
800-944-4776 / www.cincyzoo.org

Stan Hywet Hall and Gardens—Akron
888-836-5533 / www.stanhywet.org

Oklahoma

Butterfly Garden, Oklahoma City Zoo—Oklahoma City
405-424-3344 / www.okczoo.com

Wings of Wonder, Tulsa Zoo—Tulsa
918-669-6600 / www.tulsazoo.org

Oregon

Butterflies Forever—Astoria
503-738-3180 / www.oregonbutterflies.org

Wings of Wonder—Independence
503-838-0976 / www.wingsofwonder.us

Pennsylvania

Amazing Butterflies at Tyler Arboretum—Media
610-566-9134 / www.tylerarboretum.org

Butterfly Exhibit, Academy of Natural Sciences—Philadelphia
215-299-1048 / www.acnatsci.org

Butterfly Forest, Phipps Conservatory and Botanical
Gardens—Pittsburgh
412-622-6914 / www.conservatory.org

Hershey Gardens Butterfly House—Hershey
717-534-3492 / www.hersheygardens.org

Rhode Island

Newport Butterfly Zoo—Middletown
401-849-9519 / www.butterflyzoo.com

South Carolina

Butterfly House, Cypress Gardens—Moncks Corner
843-553-0515 / www.cypressgardens.info

South Dakota

Butterfly Garden, Outdoor Campus, South Dakota Game, Fish, and
Parks—Sioux Falls
605-362-2777 / www.outdoorcampus.org

Sertoma Butterfly House—Sioux Falls
605-334-9466 / www.sertomabutterflyhouse.org

Tennessee

Ijams Nature Center—Knoxville
865-577-4717 / www.ijams.org

Butterflies: In Living Color, Memphis Zoo—Memphis
901-725-3400 / www.memphiszoo.org

Texas

Ann and O. J. Weber Butterfly Garden, Lady Bird Johnson Wildflower
Center—Austin
512-292-4100 / www.wildflower.org

Butterflies: Caterpillar Flight School, San Antonio Zoo—San Antonio
210-734-7184 / www.sazoo-aq.org

Butterfly Haus, Wildseed Farms—Fredericksburg
800-848-0078 / www.wildseedfarms.com

Cockrell Butterfly Center, Houston Museum of Natural
Science—Houston
713-639-4629 / www.hmns.org

Moody Gardens—Galveston
409-683-4102 / www.moodygardens.com

River Bend Nature Center—Wichita Falls
940-767-0843 / www.riverbendnaturecenter.org

Rosine Smith Sammons Butterfly House and Insectarium
(2008)—Dallas
214-428-7476 / www.texasdiscoverygardens.org

Utah

Butterfly Exhibit, Utah's Hogle Zoo—Salt Lake City
801-584-1715 / www.hoglezoo.org

Virginia

Bristow Butterfly Habitat, Norfolk Botanical Gardens—Norfolk
757-441-5830 / www.norfolkbotanicalgarden.org/ourcollection/
explore/bristow.shtml

Butterfly Station, Danville Science Center—Danville
434-791-5160 / www.dsc.smv.org

Washington

Bug World, Butterflies and Blooms, Woodland Park Zoo—Seattle
206-684-4800 / www.zoo.org

Insect Village and Tropical Butterfly House, Pacific Science
Center—Seattle
206-443-2001 / www.pacsci.org

Wisconsin

Mosquito Hill Nature Center—New London
920-779-6433 / www.co.outagamie.wi.us/Parks/MH_home.htm

Puelicher Butterfly Wing, Milwaukee Public Museum—Milwaukee
414-278-2728 / www.mpm.edu

Puerto Rico

Reserva Natural y Mariposario Las Limas—Guayama
787-864-6037

Mexico

El Paraiso de Las Mariposas—Mexico City
+52 55 5211 1646 / www.mariposario.org.mx

Africa

Kenya

Kipepeo Butterfly Farm—Gede
+254 42 32380 / www.kipepeo.org

Mida Butterfly Farm—Watamu
866-326-7376 / www.eco-resorts.com/Butterflies.php

Nairobi Butterfly House, African Butterfly Research
Institute—Nairobi
+254 2 884554

South Africa

Butterfly House, Butterflies for Africa—Pietermaritzburg
+27 333 871356 / www.butterflies.co.za

Butterfly World South Africa—Klapmuts
+27 21 8755628 / www.places.co.za/html/butterflyworld.html

Wing Haven—Houghton
+27 82 5746688

Asia

Hong Kong

Kadoorie Farm and Botanical Garden
+852 24837200 / www.kfbg.org.hk

The Butterfly House—Ocean Park
+852 28738601 / www.oceanpark.com.hk

India

Sunken Butterfly Garden, Mughal Garden—
Rashtrapati Bhavan
www.presidentofindia.nic.in/mughalGarden.html

Indonesia

Taman Kupu Kupu, Tabanan—Bali
+62 361 814282

Japan

Itami City Museum of Insects, Koyaike—Itami-shi Hyogo
+81 727 853582 / www.itakon.com/html/english.html

Nawa Insect Museum—Gifu City
+81 582 637291 / www.gifucvb.or.jp/en/kankou/meisyo/gifukouen.
shtml

Tama Zoo Insectarium—Tokyo
+81 042 5911611 / www.tokyo-zoo.net/english/

Malaysia

Kuala Lumpur Butterfly Park—Kuala Lumpur
+60 3 26934799 / www.virtualmalaysia.com/destination/kuala%20
lumpur%20butterfly%20park.html

Malacca Butterfly and Reptiles Sanctuary—Malacca
+60 6 2320033 / www.amazingmelaka.com/category/
attractions/animal-kingdom/

Penang Butterfly Farm—Penang
+60 4 8851253 / www.butterfly-insect.com

People's Republic of China

Expo Kunming Butterfly Garden—Kunming
+86 871 5642490 / butterfly@km.col.com.cn

Phillipines

Flora Farm Butterfly House—Quezon City, Luzon
+63 2 9204113

Singapore

Butterfly Park and Insect Kingdom—Sentosa Island
+65 2750013 / www.sentosa.com.sg/explore_sentosa/attractions/
butterflypark_insectkingdom.html

Singapore Zoological Gardens—Singapore
+65 2693411 / www.zoo.com.sg

Taiwan

Butterfly Aviary, Taipei Zoo—Taipei
+886 2 29382300 / @ zoo.gov.tw

Thailand

Phuket Butterfly Garden and Insect World—Phuket
+66 76 210861 / www.phuketbutterfly.com

Orchid and Butterfly Farm, Queen Sirikit Botanic
Garden—Chiang Mai
+66 53 298171 / www.qsbg.org/index_E.asp

Australia and New Zealand

Australia

New South Wales

Coffs Harbour Butterfly House—Bonville
+61 26 6534766 / www.butterflyhouse.com.au

Queensland

Australian Butterfly Sanctuary—Kuranda
+61 7 40937575 / www.australianbutterflies.com

Australian Insect Farm—Innisfail
+61 7 40633860 / www.insectfarm.com.au

Victoria

Melbourne Zoo—Parksville
+61 3 92859300 / www.zoo.org.au/melbourne

Western Australia

Lotteries Butterfly House, Perth Zoo—Perth
+61 8 94743551 / www.perthzoo.wa.gov.au

New Zealand

Butterfly and Orchid Garden, Dickerson Holiday Park—Tararu,
Thames
+64 7 8688080 / www.butterfly.co.nz

Butterfly Creek—Auckland
+64 9 2758880 / www.butterflycreek.co.nz

Butterfly House, West Lynn Gardens—New Lynn,
Waitakere City
+64 9 8277045 / www.westlynngarden.org.nz

Europe

Austria

Natural History Museum—Vienna
+43 1 52177577 / www.nhm-wien.ac.at

Schmetterlingehaus Im Palmenhaus—Vienna
+43 1 5338570 / www.schmetterlinghaus.at

Tiergarten Schönbrunn, Vienna Zoo—Vienna
+43 1 8779294 ext. 252 / www.zoovienna.at

Belgium

De Vlindertuin, Jardin des Papillons—Knokke-Heist
+32 50 610472 / www.vlindertuin-knokke.be

Denmark

Butterfly Park and Tropical Land—Bornholm
+45 56 492575 / www.sommerfugleparken.dk

Copenhagen Zoo—Frederiksberg
+45 36302555 / www.zoo.dk

Randers Regnskov Tropical Zoo—Randers
+45 86406933 / www.regnskoven.dk/Engelsk/forside.htm

Tropical Zoo Plants—Nørresundby
+45 23436383 / www.tropicalzooplants.dk

France

The Butterfly Farm, Parc Floral de la Source—Orleans
+33 238 493000 / www.parcfloral-lasource.fr

Chateau de Goulaine—Haute Goulaine
+33 2 40549142
http://chateau.goulaine.online.fr/index-us.html

Jardin aux Papillons—Vannes
+33 2 97406740 / www.jardinauxpapillons.com

Jardins des Papillons—Ribeauville
+33 3 89733333 / www.jardinsdespapillons.fr/demo/accueil.php

Jardin des Plantes Vivarium—Paris
+33 1 43317879
http://en.wikipedia.org/wiki/Jardin_des_Plantes

La Jungle des Papillons—Antibes
+33 4 93334949 / www.marineland.fr

La Tropique du Papillon—Elne
+33 4 6837 8377 / www.tropique-du-papillon.com

Naturospace Ecuatorial—Honfleur
+33 02 31817700 / www.naturospace.com

Papillons Exotiques—Teich
+33 557 523357 / http://lejardindespapillons.free.fr/indexGB.htm

Parc Floral de Paris, Maison Paris Nature—Vincennes
+33 43 284763 / www.parcfloraldeparis.com

Germany

Botanischer Garten Munchen—Munchen
+49 89 17861316 / www.botmuc.de

Dortmund Zoo—Dortmund
+49 231 736007 / www.zoo.dortmund.de

Idea DschungelParadies—Neuenmarkt
+49 9227 902525 / www.dschungelparadies.de

Forstzoologisches Institut—Freiburg
+49 7612 033663 / www.fzi.uni-freiburg.de

Garten der Schmetterlinge Schloss—Sayn
+49 2622 15478 / www.sayn.de

Krefelder Zoo—Krefeld
+49 2151 95520 / www.zookrefeld.de

Schmetterlingslust—Berlin
+49 30 8024021

Schmetterlingshuus—Mainau Island
+49 753 1303113 / www.mainau.de

Wilhelma Insektarium, Wilhelma Zoologisch-Botanischer
Garten—Stuttgart
www.wilhelma.de

Hungary

Budapest Zoo and Botanical Garden—Budapest
+36 1 2734900 / www.zoobudapest.com

Ireland

Straffan Butterfly Farm—County Kildare
+353 1 6271109 / http://www.straffanbutterflyfarm.com/

Italy

Butterfly Arc—Padova
+39 49 8910189 / www.butterflyarc.it

Butterfly Farm D'Abruzzo—Pescara
+39 85 95898 / www.butterflyfarmabruzzo.it

Casa Delle Farfalle di Bordano, Pavees Societa Cooperativa—Udine
+39 43 2988135 / www.casaperlefarfalle.it

Casa Delle Farfalle Monteserra—Viagrande
+39 347 2686906 / www.parcomonteserra.it

Luxembourg

Jardin des Papillons, Societe Mosellane—Grevenmacher
+352 758539 / www.papillons.lu

Netherlands

Educatieve Boerderij en Zijdemuseum—Halderberge
+31 16 5320095 / www.zijdemuseum.com

Insectarium Hortus Haren—Amsterdam
+31 20 5370053 / www.hortusharen.nl

Royal Rotterdam Zoological and Botanical Garden—Rotterdam
+31 10 4431446 / www.rotterdamzoo.nl

Vlinderparadijs Papiliorama—Havelte
+31 52 1342155 / http://home.planet.nl/~hend5115/pages/indexpag
.html

Vlinderpaviljoen, Artis Zoo—Amsterdam
+31 20 5233400 / www.artis.nl

Vlindertuin, Dierenpark, Noorder Zoo—Emmen
+31 59 1850850 / www.dierenpark-emmen.nl

Vlindertuin Berkenhof—Kwadendamme
+31 11 3649729 / www.vlindertuindeberkenhof.nl

Russia

Moscow Zoo—Moscow
+7 495 2556034 / www.moscowzoo.ru

Spain

Butterfly Park Empuriabrava, Garden Center Costa Brava—Girona
+34 972450761 / www.butterflypark.es

Jardin de las Mariposas y Insectario Tropical, Zoo de Santillana del
Mar—Cantabria
+34 942818125 / www.zoosantillanadelmar.com

Mariposario del Drago—Tenerife
+34 922815167 / www.mariposario.com

Mariposario Tropical, Parque de las Ciencias—Granada
+34 958131900 / www.parqueciencias.com/novedades/mariposario/
maripo.htm

Sweden

Butterfly House, Goteborg Horticultural Gardens—Goteborg
+46 31 7411106 / www.gotbot.se

Fjarilshuset—Stockholm/Solna
+46 8 7303981 / www.fjarilshuset.se

Switzerland

Papiliorama, Swiss Tropical Gardens—Marin, Epagnier
+41 31 7560461 / www.papiliorama.ch

United Kingdom

England
Bedford Butterfly Park, Wilden—Bedfordshire
+44 1234 772770 / www.bedford-butterflies.co.uk

Berkeley Butterfly House—Gloucestershire
+44 1453 810332 / www.ukattraction.com/heart-of-england/berkeley-butterfly-house.htm

Blenheim Palace Butterfly House, Woodstock—Oxford
+44 8700 602080 / www.blenheimpalace.com/palacepg/thepleas.htm

Buckfast Butterflies, Buckfast—Devon
+44 1364 642916 / www.ottersandbutterflies.co.uk

Butterfly and Fountain World, Wootton—Isle of Wight
+44 1983 883430 / www.butterfly-world-iow.co.uk

The Butterfly House at Cotswold Wildlife Park,
Burford—Oxfordshire
+44 1993 823006 / www.cotswoldwildlifepark.co.uk

The Butterfly House at Williamson Park, Lancaster—Lancashire
+44 1524 33318 / www.williamsonpark.com

Lakeland Wildlife Oasis, Milnthorpe—Cumbria
+44 1539 563027 / www.wildlifeoasis.co.uk

The London Butterfly House at Syon Park—Middlesex
+44 20 85607272 / www.londonbutterflyhouse.com

Long Sutton Butterfly and Wildlife Park, Spalding—Lincolnshire
+44 1406 363833 / www.butterflyandwildlifepark.co.uk

Longleat Butterfly Garden, Warminster—Wiltshire
+44 1985 844400 / www.longleat.co.uk

Mole Hall Wildlife Park and Butterfly Pavilion, Saffron
Walden—Essex
+44 1799 540400 / www.molehall.co.uk

Stratford-upon-Avon Butterfly Farm,
Stratford-upon-Avon—Warwickshire
+44 1789 299288 / www.butterflyfarm.co.uk

Tropical Butterfly House, Wildlife and Falconry Centre, Sheffield—
South Yorkshire
+44 1909 569416 / www.butterflyhouse.co.uk/index.htm

Tropical Wings World of Wildlife, South Woodham Ferrers—Essex
+44 1245 425394 / www.tropicalwings.co.uk

Tropical World at Roundhay Park, Leeds—West Yorkshire
+44 113 2661850 / www.roundhaypark.org.uk

Worldwide Butterflies and Lullingstone Silk Farm,
Sherborne—Dorset
+44 1935 474608 / www.wwb.co.uk

Northern Ireland

Seaforde Gardens and Tropical Butterfly House, Seaforde—Co. Down
+44 28 44811225 / www.seafordegardens.com

Scotland

The Arran Butterfly Farm—Isle of Arran (opening 2008)
+44 131 4776550 / www.arranbutterflyfarm.co.uk

Edinburgh Butterfly and Insect World—Edinburgh
+44 131 6634932 / www.edinburgh-butterfly-world.co.uk

Wales

Conwy Butterfly Jungle—Conwy
+44 1492 593149 / www.conwy-butterfly.co.uk

Felinwynt Rain Forest Centre—Cardigan
+44 1239 810250 / www.butterflycentre.co.uk

The Magic of Life Butterfly House—Aberystwyth
+44 1970 880928 / www.magicoflife.org

Pili Palas (Butterfly Palace)—Angelsey
+44 1248 712474 / www.pilipalas.co.uk

Plantasia—Swansea
+44 1792 474555

Channel Islands

Jersey Butterfly Centre, St. Mary—Jersey
+44 1534 481707

Central America and the Caribbean

Belize

Butterfly Farm, Chaa Creek—Cayo District, San Ignacio
877-709-8708/ +501-824-2037 / www.chaacreek.com

Butterfly Garden Center, Shipstern Nature Reserve—Sarteneja
Village
+501 4 32247 / www.shipstern.org

Green Hills Butterfly Ranch—Belmopan
+501 8 204017/ www.biological-diversity.info/greenhills.htm

Tropical Wings Nature Center—Cayo District
+501 9 32265 / www.thetrekstop.com/tropwing.htm

Costa Rica

Spirogyra Butterfly Garden—San Jose
+506 2222937 / www.infocostarica.com/butterfly/index.htm

The Butterfly Farm—La Guacima, Alejuela
+506 4380400 / www.butterflyfarm.co.cr

Monteverde Butterfly Garden—Puntarenas
+506 6455512 / www.monteverdebutterflygarden.com

French West Indies

La Ferme des Papillons—St. Martin
+5995 44 3562 / www.thebutterflyfarm.com

Guatemala

El Mariposario, The Nature Reserve at San Buenaventura de
Atitlan—Panajachel
+502 9 622059 / www.atitlan.com/resnat.htm

Mariposario Antigua—Mariscal
+502 4768073 / www.antiguacultural.com/mariposario.htm

Honduras

Tropical Butterfly Farm and Gardens—El Pino, La Ceiba
+504 422874 / www.picobonito.com/PB_files/act.html#butterfly

Netherland Antillies

The Butterfly Farm—Orangestad—Aruba
+297 5 863656 / www.thebutterflyfarm.com

South America

Argentina

Mariposas del Mundo, Museum of Entomology—Buenos Aires
+54 14 6642108 / www.mariposasdelmundo.com

Brazil

Parque das Aves—Foz do Iguaçu
+45 529 8282 / www.parquedasaves.com.br/html/vl/ing/index.htm

Columbia

El Paraiso de Las Mariposas—Medellin
www.pilosos.com.co/web/diver/mascotas/zoologicos/zoomedellin/
exihibiciones.htm

Ecuador

La Selva Jungle Lodge Butterfly Farm—Quito
+593 2 2545425/ +593 2 550995 / www.laselvajunglelodge.com

Peru

The Butterfly House and Tambopata Butterfly Farm—Madre de Dios
+51 84 573534

Website Resources

In addition to the online resources listed below, see also the websites of butterfly and moth organizations detailed in Appendix E.

General Information

www.back-garden-moths.co.uk
Back Garden Moths—database and information on British moths.

www.butterfliesandmoths.org
A searchable database of butterfly and moth records in the United States and Mexico.

www.butterflybreeders.org
International Butterfly Breeders Association—lists members and information on farming.

www.butterflywebsite.com
The Butterfly Website—links to a wide variety of resources.

www.chebucto.ns.ca/Environment/NHR/lepidoptera.html
Electronic Resources on Lepidoptera—find links to many websites with varied information on butterflies and moths.

www.lepidopterology.com
Website focused on butterfly and moth studies—includes links, glossary, and bookshop.

www.leptree.net
An interactive web tool for lepidopterists to exchange information.

www.tolweb.org/tree
Tree of Life web project site—provides information about the diversity of organisms on earth, their evolutionary history, and characteristics.

www.ukbutterflies.co.uk
UK Butterfly website—provides information on butterfly species in
the United Kingdom and Ireland.

www.ukleps.org
Information and images of the eggs, larvae, and pupae of
butterflies and moths in the United Kingdom and
northern Europe.

www.virtualmuseum.ca/Exhibitions/Butterflies
Butterflies North and South—website can be viewed in English,
French, or Spanish, information on butterflies in
Canada and Peru.

Monarch Butterflies

www.learner.org/jnorth
Journey North—A global study of wildlife migration and change. Find
out about Monarch migration through news updates and maps. See
also Journey North for Kids.

http://www.monarchlab.umn.edu/default.aspx
University of Minnesota Monarch Lab—information on biology and
research.

www.worldwildlife.org/monarchs
Facts about Monarch butterflies from the
World Wildlife Fund.

Conservation

www.butterfly-conservation.org
Butterfly Conservation—a British organization committed to saving
butterflies, moths, and their habitats.

www.cites.org
Convention on International Trade in Endangered Species of Wild
Flora and Fauna (CITES)—searchable database of listed species,
information on how CITES works.

http://www.fws.gov/endangered
U.S. Fish and Wildlife Service—Endangered Species Program, lists
threatened and endangered species.

www.iucn.org
The World Conservation Union—searchable database of worldwide
threatened and endangered species.

Butterfly Gardening

www.abnativeplants.com
American Beauties—supplier of native North American plants, lots of information on the plants and the type of wildlife they attract, including nectar and host plants.

www.blossomswap.com
Blossom Swap—information and seed exchange site.

http://butterflygardeners.com
Butterfly Gardeners' Quarterly—information on butterfly gardening, links to resources and organizations.

www.gardenguides.com
Garden Guides—information on plants, pests, gardening tips and techniques, gardening recipes, seeds and bulbs, gardening books, nurseries, and landscapers.

www.invasive.org
Invasive and Exotic Species—information on species invasive to the United States.

www.nhm.ac.uk/entomology/hostplants
A database of the world's Lepidopteran host plants.

www.wild-flowers.com
Website offers links to wildflower and native plant resources.

Scientific Supplies

www.bioquip.com
Bioquip Products—equipment, supplies and books for entomology, related sciences, and collecting.

Organizations

Sampling of North American Organizations

www.troplep.org/society.htm lists organizations around the world

General

American Entomological Society (1859)
c/o Academy of Natural Sciences
1900 Ben Franklin Parkway
Philadelphia, PA 19103
215-561-3978 / Fax: 215-299-1028
aes@acnatsci.org
www.ansp.org/hosted/aes/index.html

Association for Tropical Lepidoptera (1989)
P.O. Box 141210
Gainesville, FL 32614-1210
352-392-5894 / Fax: 352-373-3249
jbhatl@aol.com (Dr. J. B. Heppner)
www.troplep.org

Entomological Society of America (1906)
10001 Derekwood Lane, Suite 100
Lanham, MD 20706-4876
301-731-4535 / Fax: 301-731-4538
esa@entsoc.org
www.entsoc.org

Entomological Society of Canada (1863)
393 Winston Avenue
Ottawa, Ontario K2A 1Y8, Canada
613-725-2619/ Fax: 613-725-9349

dixonpl@agr.gc.ca (Dr. Peggy Dixon)
www.esc-sec.org

International Association of Butterfly Exhibitors (2001)
P.O. Box 240757
St. Paul, MN 55124
952-212-4757 / Fax: 952-953-9133
info@butterflyexhibitions.org
www.butterflyexhibitions.org

Lepidopterists' Society (1947)
c/o Department of Entomology, LACM
900 Exposition Boulevard
Los Angeles, CA 90007
213-744-3364
www.lepsoc.org

North American Butterfly Association (1993)
4 Delaware Road
Morristown, NJ 07960
973-285-0907
http://www.naba.org

Sonoran Arthropod Studies Institute (1986)
P.O. Box 5624
Tucson, AZ 85703-0624
520-883-3945/ Fax: 520-883-2578
Inverticap@SASIonline.org
www.sasionline.org

Young Entomologists' Society (1965)
6907 West Grand River Avenue
Lansing MI 48906-9131
Tel. and Fax: 517-886-0630
YESbugs@aol.com
www.members.aol.com/YESbugs/bugclub.html

Monarch Butterflies

Monarch Program (1990)
P.O. Box 178671
San Diego, CA 92177
760-944-7113 / Fax: 619-466-0689
monarchprg@aol.com
www.monarchprogram.org

Monarch Watch (1992)
University of Kansas
1200 Sunnyside Avenue
Lawrence, KS 66045-7534
888-TAGGING / Fax: 785-864-4852
monarch@ku.edu
www.MonarchWatch.org

Texas Monarch Watch (1993)
Texas Parks and Wildlife
3000 South IH-35, Suite 100
Austin, TX 78704
512-912-7059 / Toll Free: 800-792-1112 ext. 7059
mike.quinn@tpwd.state.tx.us
www.tpwd.state.tx.us/learning/texas_nature_trackers/monarch/
www.texasento.net/dplex.htm

Conservation/ Education

Butterfly Conservation Initiative (2001)
McGuire Center for Lepidoptera and Biodiversity
Florida Museum of Natural History
UF Cultural Plaza
308 Hull Road
P.O. Box 112710
Gainesville, FL 32611-2710
352-392-5894 ext. 233 / bfci@aza.org / www.butterflyrecovery.org

Earthwatch Institute (1971)
3 Clock Tower Place, Suite 100
Box 75
Maynard, MA 01754
978-461-0081/ Toll Free: 800-776-0188 / Fax: 978-461-2332
info@earthwatch.org
www.earthwatch.org

Xerces Society (1971)
4828 SE Hawthorne Boulevard
Portland, OR 97215
503-232-6639 / Fax: 503-233-6794
info@xerces.org
www.xerces.org

Butterfly Gardening

National Gardening Association (1973)
1100 Dorset Street
South Burlington, VT 05403
802-863-5251
www.garden.org

North American Native Plant Society (1984)
P.O. Box 84, Station D
Etobicoke, Ontario M9A 4X1 CANADA
416-631-4438
nanps@nanps.org
www.nanps.org

Wild Ones (1977)
P.O. Box 1274
Appleton, WI 54912
877-FYI-WILD / 921-730-3986
info@for-wild.org
www.for-wild.org

APPENDIX F

Suggestions for Further Reading

Attenborough, D. 2005. *Life in the Undergrowth.* New Jersey: Princeton Univ. Press.

Brock, J. P., and K. Kaufman. 2003. *Butterflies of North America.* New York: Houghton Mifflin.

Brower, L. P., ed. 1988. *Mimicry and the Evolutionary Process.* Chicago: Univ. of Chicago Press.

Carroll, S. B. 2005. *Endless Forms Most Beautiful: The New Science of Evo Devo.* New York: Norton.

Carter, D. 1992. *Eyewitness Handbook to Butterflies and Moths.* London: DK Publishing.

DeVries, P. J. 1987. *The Butterflies of Costa Rica and Their Natural History. Papilionidae, Pieridae, Nymphalidae,* vol. 1. New Jersey: Princeton Univ. Press.

Eberhard, W. G. 1985. *Sexual Selection and Animal Genitalia.* Cambridge: Harvard Univ. Press.

Emmel, T. C. 2003. *Florida's Fabulous Butterflies,* 4th ed. Tampa, Fla.: World Publications.

———. 1997. *Butterfly Gardening: Creating a Butterfly Haven in Your Garden.* New York: Friedman/Fairfax.

———. 1975. *Butterflies.* New York: Knopf.

Glassberg, J. 2004. *Butterflies of North America.* New York: Sterling.

———. 1999. *Butterflies through Binoculars: The East.* New York: Oxford Univ. Press

Glassberg, J., M. C. Minno, and J. V. Calhoun. 2000. *Butterflies through Binoculars.* New York: Oxford Univ. Press.

Gochfeld, M., and J. Burger. 1997. *Butterflies of New Jersey.* New Jersey: Rutgers Univ. Press

Gould, J. L., and C. G. Gould. 1989. *Sexual Selection.* New York: Scientific American Library.

Grimaldi, D., and M. S. Engel. 2005. *Evolution of the Insects*. New York: Cambridge Univ. Press.

Hofmann, H., and T. Marktanner. 2001. *Collins Nature Guide to Butterflies and Moths of Britain and Europe*. London: HarperCollins.

Himmelman, J. 2002. *Discovering Moths*. Camden, Maine: Down East Books.

Howe, W. H. 1963. *Our Butterflies and Moths*. North Kansas City, Mo.: True Color Publishing.

————, ed. 1975. *The Butterflies of North America*. New York: Doubleday.

Landman, W. 1999. *The Complete Encyclopedia of Butterflies*. U.K.: Grange Books.

Larsen, T. 1992. *The Butterflies of Kenya and Their Natural History*. New York: Oxford Univ. Press.

Nijhout, H. F. 1991. *The Development and Evolution of Butterfly Wing Patterns*. Washington, D.C.: Smithsonian Institution Press.

Opler, P. A. 1994. *Butterflies and Moths*. New York: Houghton Mifflin.

Pyle, R. M. 1984. *The Audubon Society Handbook for Butterfly Watchers*. New York: Charles Scribner's Sons.

————. 1999. *Chasing Monarchs: Migrating with the Butterflies of Passage*. Boston: Houghton Mifflin.

Schappert, P. 2000. *A World for Butterflies: Their Lives, Behavior and Future*. Buffalo, N.Y.: Firefly Books.

Scheer, J. 2003. *Night Visions*. New York: Prestel.

Schilthuizen, M. 2001. *Frogs, Flies, and Dandelions*. New York: Oxford Univ. Press.

Schoonhoven, L. M., J. A. van Loon, and M. Dicke. 2006. *Insect-Plant Biology*, 2nd ed. London: Oxford Univ. Press.

Scoble, M. J. 1992. *The Lepidoptera: Form, Function and Diversity*. London: Oxford Univ. Press.

Shalaway, S. 2004. *Butterflies in the Backyard*. Mechanicsburg, Pa.: Stackpole Books.

Stadler, E., M. Rowell-Rahier, and R. Baur. 1996. *Proceedings of the 9th International Symposium on Insect-Plant Relationships (Series Entomologica)*. Dordrecht, Netherlands; Boston, Mass: Kluwer Academic.

Stokes, D., L. Stokes, and E. Williams. 1991. *Stokes Butterfly Book: The Complete Guide to Butterfly Gardening, Identification, and Behavior*. Boston: Little, Brown.

Vane-Wright, R. I., and P. R. Ackery. 1989. *The Biology of Butterflies*. New Jersey: Princeton Univ. Press.

Wagner, D. L. 2005. *Princeton Field Guide to Caterpillars of Eastern North America*. New Jersey: Princeton Univ. Press.

Whalley, P. 1988. *Butterfly and Moth*. New York: Dorling Kindersley.

Winter, W. D. 2000. *Basic Techniques for Observing and Studying Moths and Butterflies*. Los Angeles: Natural History Museum.

Wright, A. B. 1993. *Peterson First Guide to Caterpillars of North America*. Boston: Houghton Mifflin.

Xerces Society. 1998. *Butterfly Gardening: Creating Summer Magic in Your Garden*. San Francisco: Sierra Club Books.

Young, M. 1997. *The Natural History of Moths*. London: T. and A. D. Poyser.

APPENDIX G

Species List of Butterflies and Moths

Common Name	Scientific Name	General Location
Admiral, Red	*Vanessa atalanta*	All continents* and Hawaii
Admiral, White	*Limenitis arthemis arthemis*	northern United States
African Peach moth	*Egybolis vaillantina*	East Africa
African Satyrid	*Bicyclus anynana*	Southern and Central Africa
Angle Shades moth	*Phlogophora meticulosa*	Europe
Apollos	*Parnassius* spp.	Central Asia, South-western China, and Europe
Apollo	*Parnassius autocrator*	Afganistan, Tajikistan
Apollo, Small	*Parnassius phoebus*	European Alps
Apple Ermine moth	*Yponomeuta malinellus*	Europe, Asia, North America
Apple Pith moth	*Blastodacna atra*	Europe, North America
Archduke	*Lexias dirtea/Lexias pardalis*	Southeast Asia, Australasia**
Arctiid moth	*Pareuchaetes pseudoinsulata*	Venezuela, Trinidad
Atlas moth	*Attacus atlas*	Southeast Asia
Azure, Spring	*Celastrina ladon*	Alaska to Mexico
Azure, Summer	*Celastrina neglecta*	Central and Mid–Eastern U.S., Canada

Common Name	Scientific Name	General Location
Bagworm moths	Psychidae	All continents*
Bagworm moth, Finnish	*Dahlica lichenella*	Europe
Belted Beauty	*Lycia zonaria*	Europe
Birdwing, Green	*Ornithoptera priamus*	Australia, New Guinea
Birdwing, Queen Alexandra's	*Ornithoptera alexandrae*	Papua New Guinea
Birdwing, Southern	*Troides minos*	India
Blue, Alcon	*Maculinea rebeli*	Europe
Blue, Eastern Tailed	*Everes comyntas*	Eastern and Southern U.S. to Costa Rica
Blue, Karner	*Lycaeides melissa samuelis*	Central and Northeastern U.S. to Mexico
Blue, Large	*Maculinea arion*	Europe, Asia
Blue, Lotis	*Lycaeides argyrognomon lotis*	possibly extinct
Blue, Miami	*Hemiargus thomasi bethunebakeri*	Florida
Blue, Silver–studded	*Plebejus argus*	Europe, Asia
Blue, Western Pygmy	*Brephidium exilis*	Western U.S. to South America
Blue, Xerces	*Glaucopsyche xerces*	extinct
Bogong moth	*Agrotis infusa*	Australia
Brahmin moth	Brahmaeidae	Europe, Asia, North Africa
Brimstone	*Gonepteryx rhamni*	Europe, North Africa, Asia
Brown, Common Evening	*Melanitis leda*	Australia, Asia
Brown, Woodland	*Lopinga achine*	Europe
Buckeye, Common	*Junonia coenia*	U.S. to Mexico, West Indies
Calpe moth	*Calyptra eustrigata*	Southeast Asia
Cankerworm moth, Fall	*Alsophilia pometaria*	North America

Common Name	Scientific Name	General Location
Carpenter Worm moth	*Chilecomadia valdiviana*	Central/South America
Case moths	Psychidae	All continents*
Case–bearing moth	*Hyposmocoma molluscivora*	Hawaii
Cave moth, Schrankia	*Protanura hawaiiensis*	Hawaii
Cecropia moth	*Hyalophora cecropia*	Eastern and Central North America
Checkerspot, Baltimore	*Euphydryas phaeton*	Eastern North America
Clearwing	*Greta morgane*	Southern U.S., Mexico
Clearwings	Ithomiinae	Central/South America
Clothes moth, Casemaking	*Tinea pellionella*	Europe, North America, Australia
Clothes moth, Webbing	*Tineola bisselliella*	All continents*
Codling moth	*Cydia pomonella*	Europe, U.S., Australia
Comma, Eastern	*Polygonia comma*	Central and Eastern U.S., Canada
Common Mormon	*Papilio polytes*	South and Southeast Asia to Phillippines, Australia
Common Rose	*Pachliopta aristolochiae*	South Asia/ Southeast Asia
Copper, American	*Lycaena phlaeas*	North America, Europe, northern Asia
Corn Earworm	*Helicoverpa zea*	U.S.
Cossid moth	*Endoxyla leucomochla*	Australia
Cottonworm moth	*Spodoptera littoralis*	U.S., Europe, North Africa
Cracker butterflies	*Hamadryas* spp.	Central/South America, Mexico
Crescent, Pearl	*Phyciodes tharos*	North America

Common Name	Scientific Name	General Location
Crow, Common Indian	*Euploea core*	India, Southeast Asia, Australia
Crow, Double Banded	*Euploea sylvester*	South/Southeast Asia
Crow, Purple	*Euploea* spp.	Southeast Asia
Dogface, California	*Zerene eurydice*	California
"Eighty" butterfly	*Diaethria candrena*	South America
Elfin, San Bruno	*Callophrys mossii bayensis*	California
Flour moth, Mediterranean	*Ephestia kuehniella*	All continents*
Fritillary, Diana	*Speyeria diana*	Southeastern U.S.
Fritillary, Great Spangled	*Speyeria cybele*	North America
Fritillary, Gulf	*Agraulis vanillae*	Southern U.S. to South America
Fritillary, Heath	*Melitaea athalia*	Europe to Asia, Japan
Fritillary, High Brown	*Argynnis adippe*	Europe, across temperate Asia to Japan
Fritillary, Marsh	*Euphydryas aurinia*	Europe to Asia
Fritillary, Meadow	*Boloria bellona*	North America
Fritillary, Mormon	*Speyeria mormonia*	Western U.S., Canada
Fritillary, Polaris	*Boloria polaris*	Northern Scandinavia
Fritillary, Regal	*Speyeria idalia*	Central to Northeastern North America
Fruit–eating moth	*Eudocima serpentifera*	Dominican Republic, Brazil, one recorded in U.S.
Fruit–eating moth	*Ophiusa tirhaca*	Europe, Asia, Australia
Glasswing, Common	*Greta oto*	Mexico to Panama
Goatweed butterfly	*Anaea andria*	Midwest/ Southern U.S.
Grain moth, Angoumois	*Sitotroga cerealella*	All continents*

Common Name	Scientific Name	General Location
Grayling	*Hipparchia semele*	Europe, Western and Northern Asia
Great Eggfly	*Hypolimnas bolina*	Southeast Asia, Australia, New Zealand
Gypsy moth	*Lymantria dispar*	All continents*
Hairstreak, Atala	*Eumaeus atala*	Florida
Hairstreak, Banded	*Satyrium calanus*	North America
Hairstreak, Brown	*Thecla betulae*	Europe, across Asia to Korea
Hairstreak, Colorado	*Hypaurotis crysalus*	Rocky Mountains states
Hairstreak, Coral	*Satyrium titus*	North America
Hairstreak, Great Purple	*Atlides halesus*	U.S. south to Guatemala
Harvester	*Feniseca tarquinius*	Eastern North America
Hawk moth, Death's Head	*Acherontia atropos*	Africa, Europe
Hawk moth, Giant	*Xanthopan morganii praedicta*	Madagascar
Horneater moth	*Ceratophaga vicinella*	Florida
Inchworm moth	*Semiothisa bisignata*	U.S.
Indian Leafwing	*Kallima inachus*	India, Southeast Asia
Io moth	*Automeris io*	Eastern and Central North America to Costa Rica
Jay, Common	*Graphium doson*	Southeast Asia
Jay, Tailed	*Graphium agamemnon*	Southeast Asia, Australia
Jumping Bean moth	*Cydia saltitans/ Carpocapsa saltitans*	Mexico, California
Jumping Bean moth	*Laspeyresia saltitans*	Southern U.S., Mexico
Large Tree Nymph	*Idea leuconoe*	Southeast Asia
Longwings	*Heliconius* spp.	Central and South America, Southern U.S.
Longwing, Cydno	*Heliconius cydno*	Mexico to Columbia, Ecuador

Common Name	Scientific Name	General Location
Longwing, Hecale	*Heliconius hecale*	Mexico to Peru
Longwing, Zebra	*Heliconius charitonius*	Southern U.S., Central/South America, West Indies
Looper moth	*Erannis tiliaria*	U.S.
Looper moth, Blackberry	*Chlorochlamys chloroleucaria*	North America
Luna moth	*Actias luna*	Eastern North America
Malachite	*Siproeta stelenes*	Southern U.S. to Amazon
Map	*Araschnia levana*	Central Europe through Central Asia to Japan
Marion moth	*Pringleophaga marioni*	Sub–Antarctic
Metalmark moth	*Brenthia* spp.	All continents*
Mill moth	*Ephestia kuehniella*	All continents*
Monarch	*Danaus plexippus*	Canada to South America, Australia to Philippines
Mopane moth	*Imbrasia belina*	Southern Africa
Morpho, Blue	*Morpho* spp.	Central/South America
Morpho, Blue	*Morpho menelaus*	Central/South America
Morpho, Blue	*Morpho peleides*	Central/South America
Morpho, Blue	*Morpho rhetenor*	Central/South America
Mourning Cloak	*Nymphalis antiopa*	Eurasia, North America
Olivewing	*Nessaea aglaura*	Southern Mexico to Columbia, Central America
Oso Flaco moth	*Areniscythris brachypteris*	California
Owl butterfly	*Caligo* spp.	Mexico to Amazon
Owl butterfly	*Caligo atreus*	Mexico to Peru

Common Name	Scientific Name	General Location
Owl butterfly	*Caligo eurilochus*	Guatemala to Amazon
Owlet moth, Great	*Thysania agrippina*	Southern Mexico, Central America
Paper Kite	*Idea leuconoe*	Southeast Asia
Painted Lady	*Vanessa cardui*	All continents*
Peacock	*Inachis io*	Europe, Asia
Polyphemus moth	*Antheraea polyphemus*	Eastern and Central North America
Postman	*Heliconius erato / Heliconius melpomene*	Mexico to Amazon
Promethea moth	*Callosamia promethea*	Eastern North America
Purple Emperor	*Apatura iris*	Central Europe, eastward to Korea
Puss moth	*Megalopyge opercularis*	Southern U.S., Mexico
Queen	*Danaus gilippus*	Southern U.S. to Panama
Question Mark	*Polygonia interrogationis*	North America
Redbanded Leafroller moth	*Argyrotaenia velutinana*	U.S., Canada
Red–spotted Purple	*Limenitis arthemis astyanax*	Eastern U.S. to Rocky Mountains
Ringlet	*Aphantopus hyperantus*	Europe, across Asia to Japan
Royal Walnut moth	*Citheronia regalis*	Eastern North America
Rusty–tipped Page	*Siproeta epaphus*	Mexico to Peru
Silkworm moth	*Bombyx mori*	Domesticated, extinct in the wild
Silkworm moths, Giant	Saturniidae	All continents*
Silverspot, Oregon	*Speyeria zerene hippolyta*	Oregon
Six Continent	*Hypolimnas misippus*	Tropics, all continents*
Skipper, Australian Regent	*Euschemon rafflesia*	Australia

Common Name	Scientific Name	General Location
Skipper, Banana	*Erionota thrax*	Maylasia, Australasia,** Hawaii
Skipper, Crystal	*Atrytonopsis* new species 1	North Carolina Coast
Skipper, Fiery	*Hylephila phyleus*	Argentina to Eastern and Central U.S.
Skipper, Long–tailed	*Urbanus proteus*	U.S. to Mexico
Skipper, Sachem	*Atalopedes campestris*	Brazil to U.S.
Skipper, Yucca Giant	*Megathymus yuccae*	Southern U.S., Mexico
Snout	*Libytheana carinenta*	North America to Argentina, West Indies
Spanworm moth	*Operophtera bruceata*	Southern Canada, Northern U.S.
Speckled Wood	*Pararge aegeria*	Europe, North Africa
Sphinx moth, Abbott's	*Sphecodina abbottii*	Southern Canada, Eastern U.S.
Sphinx moth, Achemon	*Eumorpha achemon*	North America
Sphinx moth, Banded	*Eumorpha fasciata*	Southeastern U.S.
Sphinx moth, Blackburn's	*Manduca blackburni*	Maui, Hawaii
Sphinx moth, Tersa	*Xylophanes tersa*	Canada to Argentina
Sphinx moth, Walnut	*Amorpha juglandis*	Eastern/Central U.S., Northern Mexico
Sphinx moth, White–lined	*Hyles lineata*	North and Central America, Caribbean
Stinky Leafwing	*Historis odius*	Southern U.S. to South America
Sulphur, Clouded	*Colias philodice*	North America
Sulphur, Lyside	*Kricogonia lysides*	Southern U.S.
Sulphur, Orange	*Colias eurytheme*	North America
Swallowtail, Black	*Papilio polyxenes*	Eastern North America to Northern South America

Common Name	Scientific Name	General Location
Swallowtail, Eastern Tiger	*Papilio* or *Pterourus glaucus*	Eastern U.S.
Swallowtail, Giant	*Papilio cresphontes*	Eastern North America to South America
Swallowtail, Jamaican Giant	*Papilio homerus*	Jamaica
Swallowtail, Old World	*Papilio machaon*	Asia, Europe, Western North America
Swallowtail, Orchard	*Papilio aegeus*	Eastern Australia
Swallowtail, Oregon	*Papilio oregonius*	Northwestern North America
Swallowtail, Pipevine	*Battus philenor*	North America
Swallowtail, Polydamas	*Battus polydamas*	Florida, South Texas to Argentina
Swallowtail, Scarce	*Iphiclides podalirius*	Europe
Swallowtail, Schaus	*Heraclides aristodemus ponceanus*	South Florida
Swallowtail, Spicebush	*Papilio troilus*	Eastern North America
Swallowtail, Two–tailed	*Papilio multicaudata*	Western North America
Swallowtail, Ulysses	*Papilio ulysses*	Australia
Swallowtail, Zebra	*Eurytides marcellus*	Central and Eastern U.S.
Tapestry moth	*Trichophaga tapetzella*	All continents*
Tear–feeding moth	*Hemiceratoides hieroglyphica*	Madagascar
Tear–feeding moth	*Lobocraspis griseifusa*	Southeast Asia
Tiger moths	Arctiidae	All continents*
Tiger, Dark Blue	*Tirumala septentrionis*	South/Southeast Asia
Tigerwing, Olivencia	*Forbestra olivencia*	Amazon basin
Tortoiseshell, Compton	*Nymphalis vaualbum*	Subarctic Alaska to Central U.S.
Tortoiseshell, Large	*Nymphalis polychloros*	Central/Southern Europe, North Africa, Asia

Common Name	Scientific Name	General Location
Tussock moth	*Orgyia* spp.	North Africa, Asia, Europe, North America
Vampire moth	*Calyptra eustrigata*	Southeast Asia
Vapourer moth	*Orgyia antiqua*	U.K.
Viceroy	*Limenitis archippus*	Canada to Northern Mexico
Water Veneer moth	*Acentria ephemerella*	Europe and North America
Wax moth, Greater	*Galleria mellonella*	All continents*
Wax moth, Lesser	*Achroia grisella*	All continents*
White, Cabbage	*Pieris rapae*	Europe, North Africa, Asia, North America
White, Great Southern	*Ascia monuste*	Southern U.S., Mexico
White, Large	*Pieris brassicae*	Europe, North Africa, Asia
White, Marbled	*Melanargia galathea*	Europe, North Africa
White, Western	*Pontia occidentalis*	Western North America
White, Yellow Patch	*Colotis halimede*	Senegal, East Africa, Saudi Arabia
White Peacock	*Anartia jatrophae*	Southern U.S., Central/South America, West Indies
Witch moth, Black	*Ascalapha odorata*	Southern U.S. to Brazil, Caribbean
Witch moth, White	*Thysania agrippina*	Southern Mexico, Central America
Wood Nymph	*Redonda bordoni*	Northern South America
Yellow, Clouded	*Colias crocea*	Europe, North Africa
Yellow Tip	*Anthocharis scolymus*	Eastern Asia
Yucca moth	*Tegeticula* spp.	Southern U.S.
No Common Names	*Palla ussheri*	Central Africa
	Pringleophaga kerguelensis	Sub–Antarctic

* There are no butterflies in Antartica.
** Australasia is a distinct region with a common evolutionary history and many unique plants and animals sharing a common ancestry. It includes Australia, New Zealand, New Guinea, the eastern part of the Indonesian archipelago, and several nearby Pacific island groups.

Sources

Chapter 1

Crabbe, N. "UF Researchers Work to Help Tiny Butterfly Make a Comeback." On the Web site Friends of Sand Mountain. http://www.sandmountain-nv.org/index.php?name=PNphpBB2&file=viewtopic&p=9774. Accessed February 18, 2008.

Chapter 2

Scott, J. A. 1986. *The Butterflies of North America: A Natural History and Field Guide*. Palo Alto, California: Stanford University Press.

Yack, J. E. et al. 2000. "Sound Production and Hearing in the Blue Cracker Butterfly *Hamadryas feronia* (Lepidoptera, nymphalidae) from Venezuela." *Journal of Experimental Biology*. 203:3689–3702.

Yack, J. E., and J. H. Fullard. 2000. "Ultrasonic Hearing in Nocturnal Butterflies." *Nature*. 403: 265–266.

Chapter 3

Banziger, H. 1975. "Skin-Piercing Blood-Sucking Moths I: Ecological and Ethological Studies on Calpe estrugata (Lepid., Noctuidae)." *Acta Tropica*. 32:125–144.

Banziger, H. 1979. "Skin-Piercing Blood-Sucking Moths II: Studies on a Further 3 Adult Calyptra (Calpe) sp. (Lepid., Noctuidae)." *Acta Tropica*. 36:23–37.

Calvert, B. "How High Do Monarch Butterflies Fly During Fall Migration?"

On the Web site Monarch Butterfly Journey North. http://www.learner.org/jnorth/tm/monarch/HeightFallFlight.html. Accessed February 18, 2008.

Gibo, D. "The Challenge of Butterfly Migration." On the Web site Tactics and Vectors. http://www.utm.utoronto.ca/~w3gibo/the.htm. Accessed February 18, 2008.

Hay-Roe, M., and R. W. Mankin. 2004. "Wing-Click Sounds of Heliconius cydno alithea (Nymphalidae: Heliconiinae) Butterflies." *Journal of Insect Behavior.* 17:329–335.

Hilgartner, R. et al. 2007. "Malagasy Birds as Hosts for Eye-Frequenting Moths." *Biology Letters.* 3:117–120. [My paper]

Kandori, I., and N. Ohsaki. 1996. "The Learning Abilities of the White Cabbage Butterfly, *Pieris Rapae,* Foraging For Flowers." *Researches on Population Ecology.* 38:111–117.

Larsen, T.B. 1983. "Strange Bedfellows." *Saudi Aramco World.* November/December.2–3.

Mallet, J. et al. 2007. "Natural Hybridization in Heliconiine Butterflies: The Species Boundary as a Continuum." *BMC Evolutionary Biology.* 7:28.

Nice, G. C., and A. M. Shapiro. 1999. "Molecular and Morphological Divergence in the Butterfly Genus *Lycaeides* (Lepidoptera: Lycaenidae) in North America: Evidence of Recent Speciation." *Journal of Evolutionary Biology.* 12:936–950.

Opler, P. A., and A. B. Wright. 1999. *Peterson Field Guide to Western Butterflies.* Boston: Houghton Mifflin.

Rutowski, R. L. 1998. "Mating Strategies in Butterflies." *Scientific American.* 279:64–69.

Viloria, A. L. et al. 2003. "A Brachypterous Butterfly?" *Proceedings of the Royal Society: Biological Sciences, Biology Letters Supplement 1.* 270:21–24.

Weiss, M. R. 1997. "Innate Colour Preferences and Flexible Colour Learning in the Pipevine Swallowtail." *Animal Behaviour.* 53:1043–1052.

Chapter 4

Alexander, R. D. 1964. "The Evolution of Mating Behaviour in Arthropods." *Symposia of the Royal Entomological Society of London.* 2:78–94.

Brakefield, P. M. et al. 1996. "Development, Plasticity and Evolution of Butterfly Eyespot Patterns." *Nature.* 384:236–242.

Cook, C. E., Q. Yue, and M. Akam. 2005. "Mitochondrial Genomes Suggest That Hexapods and Crustaceans are Mutually Paraphyletic." *Proceedings of the Royal Society: Biological Sciences.* 272:1295–1304.

Eberhard, W. G. 1985. *Sexual Selection and Animal Genitalia.* Cambridge: Harvard University Press.

Fisher, R. A., and J. H. Bennett. 2000. *The Genetical Theory of Natural Selection.* Oxford: Oxford University Press.

Glenner, H. et al. 2006. "Evolution: The Origin of Insects." *Science.* 314:1883–1884.

Gompert, Z. et al. 2006. "Homoploid Hybrid Speciation in an Extreme Habitat." *Science.* 314:1923–1925.

Janteur, P. 2005. "Epigenetics and Chromatin." *Proceedings of the Royal Society: Biological Sciences.* 272:1295–1304.

Lukhtanov, V. A. et al. 2005. "Reinforcement of Pre-Zygotic Isolation and Karyotype Evolution in *Agrodiaetus* Butterflies." *Nature.* 436:385–389.

Mallet, J. et al. 1998. "Mimicry and Warning Color at the Boundary Between Races and Species." In: *Endless Forms: Species and Speciation,* eds. D. Howard and S. Berlochers, 390–403. Oxford: Oxford University Press.

Mavarez, J. et al. 2006. "Speciation by Hybridization in Heliconius Butterflies." *Nature.* 441:868–871.

Merrifield, F. 1890. "Systematic Temperature Experiments on Some Lepidoptera in All Their Stages." *Transactions of the Entomological Society of London.* 131–159.

Merrifield, F. 1893. "The Effects of Temperature in the Pupal Stage on the Colouring of *Pieris napi, Vanessa atalanta, Chrysophanus phloeas, and Ephyra punctaria.*" *Transactions of the Entomological Society of London.* 425–438.

Nice, C., and J. Fordyce. 2006. "How Caterpillars Avoid Overheating: Behavioral and Phenotypic Plasticity of Pipevine Swallowtail Larvae." *Oecologia.* 146:541–548.

Nijhout, H. F. 1984. "Color Pattern Modification by Cold-Shock in Lepidoptera." *Journal of Embryology and Experimental Morphology.* 81:287–305.

Ramos, D. M. et al. 2006. "Temporal and Spatial Control of Transgene Expression Using Laser Induction of the *Hsp70* Promoter." *BMC Developmental Biology.* 6:55

Riddihough, G., and E. Pennisi. 2001. "The Evolution of Epigenetics." *Science.* 293:1063.

Schilthuizen, M. 2003. "Shape Matters: The Evolution of Insect Genitalia." *Proceedings of the Section Applied and Experimental Entomology of the Netherlands Entomological Society.* 14:9–15.

West-Eberhard, M. J. 1989. "Phenotypic Plasticity and the Origins of Diversity." *Annual Review of Ecology and Systematics.* 20:249–278.

Chapter 5

Boggs, C. L., and C. L. Ross. 1993. "The Effect of Adult Food Limitation on Life History Traits in Speyeria Mormonia (Lepidoptera: Nymphalidae)" *Ecology.* 74:433–441.

Cordero, C., and W. Eberhard. 2002. "Female Choice of Sexually Antagonistic Male Adaptations: A Critical Review of Some Current Research." *Journal of Evolutionary Biology.* 16:1–6.

Fischer, K. et al. 2003. "Cooler Butterflies Lay Larger Eggs: Developmental Plasticity versus Acclimation." *Proceedings: Biological Sciences.* 270:2051–2056.

Gilbert, L. E. 1976. "Postmating Female Odor in Heliconius Butterflies: A Male-Contributed Antiaphrodisiac?" *Science.* 193:419–420.

Grondahl, C. 1996. "A Closer Look: Giant moths of North Dakota." *North Dakota Outdoors.* 59:10–11

Hill, R. I. 2006. "Life History and Biology of Forbestra olivencia (Bates, 1862) (Nymphalidae, Ithominae)." *Journal of the Lepidopterists' Society.* 60:203–210.

Kopper, B. J., R. E. Charlton, and D. C. Margolies. 2004. "Oviposition Site Selection by the Regal Fritillary, Speyeria idalia, as Affected by Proximity of Violet Host Plants." *Journal of Insect Behavior.* 13:651–665.

Kopper, B.J. et al. 2001. "Evidence for Reproductive Diapause in the Fritillary *Speyeria idalia* (Lepidoptera: Nymphalidae)". *Annals of the Entomological Society of America.* 94:427–432.

Mallet, J. et al. 2007. *"Natural Hybridization In Heliconiine Butterflies: The Species Boundary As A Continuum."* BMC Evolutionary Biology. 7:28.

Mitchell, M. R. 1975. "Disruption of Pheromonal Communication Among Coexistent Pest Insects with Multichemical Formulations." *BioScience.* 25:493–499.

Myers, J. 1972. "Pheromones and Courtship Behavior in Butterflies." *American Zoologist.* 12:545–551.

Silvegren G., C. Löfstedt, and W. Q. Rosén. 2005. "Circadian Mating Activity and Effect of Pheromone Pre-exposure on Pheromone Response Rhythms in the Moth *Spodoptera littoralis.*" *Journal of Insect Physiology.* 51:277–286.

Thomas, J. A., and R. Lewington. 1991. *The Butterflies of Britain and Ireland.* London: Dorling Kindersley.

Thomas, J. A., and E. A. Maitland.1990. "The Brown Hairstreak." In: *The Butterflies of Great Britain and Ireland, Vol 7, Part1.* eds. A. M. Emmet and J. Heath. Colchester: Harley Books.

Wunderer, H. et al. 1986. "Sex Pheromones of Two Asian Moths (Creatonotos Transiens, C. gangis; Lepidoptera—Arctiidae): Behavior, Morphology, Chemistry, and Electrophysiology." *Experimental Biology.* 46:11–27.

Chapter 6

Clench, H. K. 1966. "Behavioral Thermoregulation in Butterflies." *Ecology.* 47:1021–34.

Coombs, E. M. et al., eds. 2004. *Biological Control of Invasive Plants in the United States.* Corvallis: Oregon State University Press.

Danks, H. V. 2004. "Seasonal Adaptations in Arctic Insects." *Integrative and Comparative Biology.* 44:85–94.

Deyrup, M. et al. 2005. "A Caterpillar that Eats Tortoise Shells." *American Entomologist.* 51:245–248.

Downes, J. A. 1964. "Arctic Insects and Their Environment." *Canadian Entomologist.* 96:279–307.

Jogar, K. et al. 2005. "Physiology of Diapause in Pupae of *Pieris brassicae L.* (Lepidoptera: Pieridae)." *Agronomy Research.* 3:21–37.

Kevan, P. G., T. S. Jensen, and J. D. Shorthouse. 1982. "Body Temperatures and Behavioral Thermoregulation of High Arctic Woolly-Bear Caterpillars and Pupae (Gynaephora rossii, Lymantriidae: Lepidoptera) and the Importance of Sunshine." *Arctic and Alpine Research.* 14: 125–136.

Kevan, P. G., and J. D. Shorthouse. 2003. "Behavioural Thermoregulation by High Arctic Butterflies" *Arctic.* 23:268–279.

Kukal, O., B. Heinrich, and J. G. Duman. 1988. "Behavioural Thermoregulation in the Freeze-Tolerant Arctic Caterpillar, *Gynaephora groenlandica.*" *Journal of Experimental Biology.* 138: 181–193.

Rubinoff, D., and W. P. Haines. 2005. "Web-Spinning Caterpillar Stalks Snails." *Science.* 309:575.

Youngsteadt, E., and P. J. DeVries. 2005. "The Effects of Ants on the Entomophagous Butterfly Caterpillar Feniseca tarquinius, and the Putative Role of Chemical Camouflage in the Feniseca-Ant Interaction." *Journal of Chemical Ecology.* 31: 2091–2109.

Chapter 7

Bingham, C. T. 1907 *Fauna of British India Butterflies, Volume 2.* London: Evans.

Brakefield, P. M. 1998. "The Evolution-Development Interface and Advances with the Eyespot Patterns of *Bicyclus* Butterflies." *Heredity*. 80:265–272.

Brower, L. P. 1988. "Avian Predation on the Monarch Butterfly and Its Implications for Mimicry Theory." *The American Naturalist, Supplement: Mimicry and the Evolutionary Process*. 131:S4-S6.

Brower, L., and S. Glazier. 1975. "Localization of Heart Poisons in The Monarch Butterfly." *Science*. 188:19–25.

Caffrey, D. J. 1983. *Health Hazard Evaluation Report No. GHETA 81–121*. Morgantown, West Virginia: National Institute for Occupational Safety and Health.

Carroll, S. B. 2005. *Endless Forms Most Beautiful: The New Science of Evo Devo*. New York: W.W. Norton.

Charlat, S. et al. 2005. "Prevalence and Penetrance Variation of Male-Killing *Wolbachia* Across Indo-Pacific Populations of the Butterfly *Hypolimnas bolina*." *Molecular Ecology*.14:3525–3530.

Daniels, J. C., and S. J. Sanchez. 2006. "Blues' Revival: Can a Change in Diet—and a Little Laboratory Assistance—Help a Florida Butterfly Escape Extinction." *Natural History*. 26–28 October.

Davidson, E. H. 2006. *The Regulatory Genome: Gene Regulatory Networks in Development and Evolution*. San Diego: Academic Press/Elsevier.

Davidson, E. H., and D. H. Erwin. 2006. "Gene Regulatory Networks and the Evolution of Animal Body Plans." *Science*. 311:796–800.

Eagleman, D. M. 2007. "Envenomation by the Asp Caterpillar (*Megalopyge opercularis*)." *Clinical Toxicology*. 99999:1–5.

Engler, H. S., K. C. Spencer, and L. E. Gilbert. 2000. "Preventing Cyanide Release from Leaves." *Nature*. 406:144–145.

Fatouros, N. E. et al. 2005. "Chemical Communication: Butterfly Anti-Aphrodisiac Lures Parasitic Wasps." *Nature*. 433:704.

Foot, N. C. 1922. "Pathology of the Dermatitis Caused by Megalopyge Opercularis, a Texan Caterpillar." *Journal of Experimental Medicine*. 35:737–753.

Glendinning, J. I., A. A. Mejia, and L. P. Brower. 1988. "Behavioral And Ecological Interactions of Foraging Mice (*Peromyscus melanotis*) with Overwintering Monarch Butterflies (*Danaus plexippus*) in México." *Oecologia*. 75:222–227.

Hoy, R. R. 1992. "The Evolution of Hearing in Insects as an Adaptation to Predation by Bats." In: *The Evolutionary Biology of Hearing*. eds. D.B. Webster, R.R. Fay, and A.N. Popper. 115–129. New York: Springer.

Kemp, D. J. 2003 "Twilight Fighting in the Evening Brown Butterfly *Melanitis leda* (L.) (Nymphalidae): Age and Residency Effects." *Behavioral Ecology and Sociobiology*. 54:7–13.

Kemp, D.J., and C. Wiklund. 2001. "Fighting Without Weaponry: A Review of Male-Male contest Competition in Butterflies." *Behavioral Ecology and Sociobiology*. 49:429–442.

Kemp, D. J., C. Wiklund, and K. Gotthard. 2006. "Life History Effects Upon Contest Behaviour: Age as a Predictor of Territorial Contest Dynamics in Two Populations of the Speckled Wood Butterfly, *Pararge aegeria L.*" *Ethology*. 112:471–477.

Kemp, D. J., C. Wiklund, and H. Van Dyck. 2006. "Contest Behaviour in the Speckled Wood Butterfly (*Pararge aegeria*): Seasonal Phenotypic Plasticity and the Functional Significance of Flight Performance." *Behavioral Ecology and Sociobiology*. 59:403–411.

Kinoshita, M. 1998. "Effects Of Time-Dependent Intraspecific Competition on Offspring Survival in the Butterfly, *Anthocharis scolymus* (L.) (Lepidoptera: Pieridae)." *Oecologia*. 114:31–36.

Makinen-Kiljunen, S. et al. 2001. "A Baker's Occupational Allergy to Flour Moth (Ephestia Kuehniella)." *Allergy*. 56:696–700.

Nijhout, H. F. 1980. "Pattern Formation on Lepidopteran Wings: Determination of an Eyespot." *Developmental Biology*. 80:267–274.

Nijhout, H. F. 1991. *The Development and Evolution of Butterfly Wing Patterns*. Washington, D.C.: Smithsonian Institution Press.

Orians, G., and D. Janzen. 1974. "Why are embryos so tasty?" *The American Naturalist*. 108:581–592.

Popa, V., S. George, and O. Gavanescu. 1970. "Occupational and Non-Occupational Respiratory Allergy in Bakers." *Allergy*. 25:159–177.

Takeuchi, T. 2006. "Matter of Size or Matter of Residency Experience? Territorial Contest in a Green Hairstreak, *Chrysozephyrus smaragdinus* (Lepidoptera: Lycaenidae)." *Ethology*. 112:293–299.

Thomas, J. A. et al. 2002. "Insect Communication: Parasitoid Secretions Provoke Ant Warfare." *Nature*. 417:505–506.

Wagner, D. L. 2005. *Princeton Field Guide to Caterpillars of Eastern North America*. New Jersey: Princeton University Press.

Wagner, D. L., and J. Rota. 2006. "Evasive Prey Mimicry in a Tropical Tortricid?" *News of the Lepidopterists' Society*. 48:115.

Chapter 8

Betalden, R. V., K. S. Oberhauser, and A. T. Peterson. 2007. "Ecological Niches in Sequential Generations of Eastern North American Monarch Butterflies (Lepidoptera: Danaidae): The Ecology of

Migration and Likely Climate Change Implications." *Environmental Entomology*. 36: 1365–1373.

Dunlap, J. C. 1999. "Molecular Bases for Circadian Clocks." *Cell*. 96:271–290.

Gibo, D. L., and M. J. Pallett. 1979. "Soaring Flight of Monarch Butterflies, *Danaus plexippus* (Lepidoptera: Danaidae) During the Late Summer Migration in Southern Ontario." *Canadian Journal of Zoology*. 57:1393–1401.

Giebultowicz, J. M. 2000. "Molecular Mechanism and Cellular Distribution of Insect Circadian Clocks." *Annual Review of Entomology*. 45:769–793.

Goehring, E. 1999. "Environmental Factors Involved in Reproductive Diapause in Monarch Butterflies, *Danaus plexippus*." *Master's Thesis*: University of Minnesota.

Goehring, E. "How Do Monarchs Know When to Leave?" On the Web site Monarch Lab. www.monarchlab.umn.edu/Research/Mig/migback1.html. Accessed February 18, 2008.

Hsiao, E. 2007. "Government Bureau Mulls Over Measures to Protect Migratory Butterflies from Traffic." *Taiwan Journal*. 25.

Kishen Das, K. R. on the Web site Flutter-by Migration. http://butterflyindianews.blogspot.com/2005/12/flutter-by-migration.html. Accessed February 18, 2008.

Larsen, T. 1992. *The Butterflies of Kenya and Their Natural History*. New York: Oxford University Press.

Mouritsen, H., and B. J. Frost. 2002. "Virtual Migration in Tethered Flying Monarch Butterflies Reveals Their Orientation Mechanisms."
Proceedings of the National Academy of Sciences of the United States of America. 99:10162–10166.

Pyle, R. M. 1999. *Chasing Monarchs: Migrating with the Butterflies of Passage*. Boston: Houghton Mifflin.

Reppert, S., H. Zhu, and R. White. 2004. "Polarized Light Helps Monarch Butterflies Navigate." *Current Biology*. 14:155–158.

Salvato, M. H., J. V. Calhoun, and H. L. Salvato. 2006. "Observations of Kricogonia Lyside (Pieridae) in the Florida Keys." *Journal of the Lepidopterists' Society*. 60:172–174.

Taylor, O. R. "Research Projects: Size and Mass." On the Web site Monarch Watch. http://www.monarchwatch.org/class/studproj/mass.htm. Accessed February 18, 2008.

Urquhart, F. A. 1998. *The Monarch Butterfly: International Traveler*. London: William Caxton Ltd.

Wassenaar, L., and K. Hobson. 1998. "Natal Origins of Migratory Monarch Butterflies at Wintering Colonies in Mexico: New Isotopic Evidence." *Proceedings of the National Academy of Sciences.* 95:15436–15439.

Zhu, H. et al. 2008. "Cryptochromes Define a Novel Circadian Clock Mechanism in Monarch Butterflies That May Underlie Sun Compass Navigation." *Public Library of Science: Biology.* 6:138–155.

Chapter 9

Audubon, J. J. 1937. *The Birds of America.* New York: Macmillan.

Black, S. H., and D. M. Vaughan. 2005. "Species Profile: *Lycaeides idas lotis.*" In: *Red List of Pollinator Insects of North America.* eds. M. D. Shepherd, D. M. Vaughan, and S. H. Black. Portland, Oregon: The Xerces Society for Invertebrate Conservation.

Coastwatch Spring 2007. "Naturalist's Notebook: Bogue's New Butterfly." On the Web site Sea Grant North Carolina. www.ncseagrant .org/index.cfm?fuseaction=story&pubid=147&storyid=247 Accessed February 18, 2008.

Cranshaw, W. S. "Insect Control: Soaps and Detergents." On the Web site Colorado State University: Horticulture. www.ext.colostate .edu/PUBS/insect/05547.html. Accessed February 18, 2008.

Daniels, J. C., and S. J. Sanchez. 2006. "Blues' Revival: Can a Change in Diet—and a Little Laboratory Assistance—Help a Florida Butterfly Escape Extinction." *Natural History.* 26–28 October.

Davis, C. J., and K. Kawamura. 1975. "Notes and exhibitions: *Erionota thrax* L. Proc." *Hawaiian Entomological Society.* 22:21.

Glassberg, J. 1999. *Butterflies through Binoculars: The East.* New York: Oxford University Press.

Glassberg, J. "Saving South Florida's Butterflies:Miami Blue Fund." On the Web site North American Butterfly Association. www.naba .org/miamiblue.html. Accessed February 18, 2008.

Glassberg, J. et al. "There's No Need to Release Butterflies—They're Already Free." On the Web site North American Butterfly Association. http://www.naba.org/weddings.html. Accessed February 18, 2008.

Kinver, M. "Successful Summer for Large Blue." On the Web site BBC News. http://news.bbc.co.uk/2/hi/science/nature/5385488.stm. Accessed February 18, 2008.

Lee, M. "Large Blue Butterfly- Reintroduction into North Cornwall." On the Web site The Wildlife Trusts: Cornwall. http://www .cornwallwildlifetrust.org.uk/nature/inverteb/largeblue.htm. Accessed February 18, 2008.

McCarthy, M. J. "Unsure About Butterfly Releases? A Position Paper by the International Butterfly Breeders Association." On the Web site of the International Butterfly Breeders Association. http://www.butterflybreeders.org/. Accessed February18, 2008.

McManus, M. et al. 1989. "Gypsy Moth: Forest Insect & Disease Leaflet 162." *U.S. Department of Agriculture Forest Service.*

Parmesan, C. et al. 1999. Poleward Shift of Butterfly Species' Ranges Associated With Regional Warming. *Nature.* 399:579–583.

Pyle, R. M. 1974. *Watching Washington Butterflies.* Seattle: Seattle Audubon Society.

Rudolph, D. C. et al. 2006. "The Diana Fritillary (*Speyeria diana*) and Great Spangled Fritillary (*S. cybele*): Dependence on Fire in the Ouachita Mountains of Arkansas." *Journal of the Lepidopterists' Society.* 60:218–226.

U.S. Fish & Wildlife Service. 2003. *Karner Blue Butterfly Recovery Plan (Lycaeides melissa samuelis).*

Vedder, L. A. 2006. *John James Audubon and The Birds of America: A Visionary Achievement in Ornithology Illustration.* San Marino, California: Huntington Library Press.

Weinzierl, R., and T. Henn. 1991. "Alternatives in Insect Management: Biological and Biorational Approaches, Publication 401." *North Central Region Extension Publication.*

Winter, W. D. 2000. *Basic Techniques for Observing and Studying Moths and Butterflies.* Los Angeles: Natural History Museum.

Young, M. 1997. *The Natural History of Moths.* London: T. and A. D. Poyser.

Chapter 10

Cunningham, A. S. 2000. *Crystal Palaces.* New York: Princeton University Press.

de Vleeschouwer, O. 2000. *Greenhouses and Conservatories.* Paris: Flammarion.

Wagner, D. L. 2005. *Princeton Field Guide to Caterpillars of Eastern North America.* New Jersey: Princeton University Press.

Index

Page numbers in italics refer to figures. Color plates in the text are designated by pl.

About the Authors

Hazel Davies coordinates the Butterfly Conservatory at the American Museum of Natural History in New York as part of her role as the Manager of Living Exhibits. As an experienced science teacher, she enjoys the opportunity to educate about butterflies on a daily basis. Hazel serves on the Board of Directors of the International Association of Butterfly Exhibitions, and regularly contributes her expertise and photographs to publications and professional organizations. Growing up in rural England, she developed an interest in nature at an early age, where her passion for wildlife and love of photography inspired her to travel throughout the Americas, Europe, and Asia.

Carol A. Butler is a psychotherapist and a mediator in private practice in New York City who is also an experienced photographer. She has worked as a docent in the Butterfly Conservatory at the American Museum of Natural History for the past three years and has participated in scientific and photographic expeditions to study butterflies in the United States, Europe, and Africa. She has exhibited her photographs in New York, and one of her photographs is featured on the cover of the January 2006 *Journal of the Lepidopterists' Society*. She is the coauthor of the well-received *Divorce Mediation Answer Book*, and she has written articles that have appeared in magazines, journals, and online.